Does life exist on other planets? This topical book presents the scientific basis for thinking there may be life elsewhere in the Universe. It is the first to cover the entire breadth of recent exciting discoveries, including the discovery of planets around other stars and the possibility of fossil life in meteorites from Mars.

Suitable for the general reader, this authoritative book avoids technical jargon and is well illustrated throughout. It covers all the major topics, including the origin and early history of life on Earth, the environmental conditions necessary for life to exist, the possibility that life might exist elsewhere in our Solar System, the occurrence of planets around other stars and their habitability, and the possibility of intelligent extraterrestrial life.

For all those interested in understanding the scientific evidence for and likelihood of extraterrestrial life, this is the most comprehensive and readable book to date.

BRUCE JAKOSKY is a Professor of Geology and a member of the Laboratory for Atmospheric and Space Physics at the University of Colorado. There, he teaches both undergraduate and graduate level courses in geology and planetary science. He has been at CU since 1982, after receiving his Ph.D degree from Caltech. He began Mars research working on the *Viking* mission to Mars in 1975 as an undergraduate at University of California at Los Angeles. Today, he is trying to understand the nature and evolution of the martian climate and its volatile history, and the connection between the geology and possible biology on Mars. He has been involved with a number of spacecraft missions, including *Clementine*, *Mars Observer*, and the *Mars Global Surveyor*.

Professor Jakosky was born in Washington D.C. and grew up in Southern California. He lives with his spouse in Superior, Colorado, where he is one of the few people who don't ski.

The Search for Life on Other Planets

Bruce Jakosky

CAMBRIDGE
UNIVERSITY PRESS

PUBLISHED BY THE PRESS SYNDICATE OF THE UNIVERSITY OF CAMBRIDGE
The Pitt Building, Trumpington Street, Cambridge CB2 1RP, United Kingdom

CAMBRIDGE UNIVERSITY PRESS
The Edinburgh Building, Cambridge CB2 2RU, UK http://www.cup.cam.ac.uk
40 West 20th Street, New York, NY 10011-4211, USA http://www/cup.org
10 Stamford Road, Oakleigh, Melbourne 3166, Australia

First published 1998

Printed in the United Kingdom at the University Press, Cambridge

Typeset in Utopia 9/13pt, in QuarkXPress™ [SE]

A catalogue record for this book is available from the British Library

Library of Congress cataloguing in publication data
Jakosky, Bruce M.
The search for life on other planets / by Bruce Jakosky.
     p.   cm.
Includes bibliographical references and index.
ISBN 0 521 59165 1. – ISBN 0 521 59837 0 (pbk.)
1. Life on other planets.   I. Title.
QB54.J25   1998
576.8'39–dc21   97-51549 CIP

ISBN 0 521 59165 1 hardback
ISBN 0 521 59837 0 paperback

Contents

Preface

The possibility that there might be life on other planets has become a major topic of public debate since 1995. The two specific incidents that triggered the discussions were the discoveries of planets around other stars and of possible fossil evidence in meteorites that have come from Mars for life there. Neither of these discoveries, however, provides compelling evidence for the existence of extraterrestrial life, and it is unlikely that the existence of life elsewhere will be confirmed before the end of the century. Despite these uncertainties many, if not most, scientists believe that it is very likely that life exists.

The sudden emergence of the topic of life on other planets as a major issue in our society really is more a matter of public awareness rather than of scientific advance or discovery. The issues related to life on other planets – the origin and early evolution of life on Earth, the environmental conditions required for the existence of life, the occurrence of these conditions on other planets in our solar system, the formation of planets and their possible existence around other stars, and the nature of intelligent life – have been of considerable scientific interest for a long time, and considerable efforts have been ongoing in each of these areas for decades. Within the last decade, especially, new advances in planetary science and in terrestrial biology have pointed toward the possibility of extraterrestrial life. It is only with the recent discoveries, though, that the issue has bubbled up again into the realm of public awareness.

Within the mainstream planetary science community, many of the issues related to life in the universe have, in turn, been shunned and embraced. In the years following the *Viking* spacecraft mission to Mars in 1976 to look for life, very few planetary scientists were working on questions pertaining to life on other planets and there was very little crossover between planetary science and exobiology or origin-of-life research. Since the mid-1990s, however, there has been much more crossover, and the biological questions have come to the forefront again. Although cynics will argue that this is merely a play for more funding and public attention, rather it appears to be due largely to the strength of the intellectual arguments that point to the possibility that life could exist elsewhere. These issues center on the ease with which life arose on Earth, the likely occurrence of the environmental conditions necessary for an origin of life elsewhere in our solar system, and the likely existence of planets orbiting other stars that also share these conditions.

Certainly, my own views have followed this evolution. When I became involved in planetary science research during the *Viking* era, it quickly became clear to me that the path to respectability lay in not pursuing exobiological issues. When I began teaching a decade later, my undergraduate course in controversial issues in planetary science focused exclusively on the geological questions. The strong connection between planetary science and biology, though, became clear with the discovery that an asteroid impact must have played a major role in the extinction of the dinosaurs (and other species) some 65 million years ago. Teaching introductory geology to first-year students helped underscore to me that the connections on Earth between the geological and the biological events of the last 4 billion years were strong. Over time, the question of life on other planets pushed its way more and more into my planetary course until, this year, it was taught under the name "Extraterrestrial life".

This book grew out of the background that was necessary in order to teach the material, combined with the lack of a suitable text. My goal in writing this book is to provide an introduction to the questions of life on other planets at a level suitable for the educated public or for an undergraduate college course for non-majors. This book also should be suitable as an introduction to the major questions and an entry into the literature for graduate students, since most graduate students will not be well versed in all of the areas touched on here.

Although the question of life in the universe could be approached from the perspective of chemistry, biology, geology, or astrophysics, I've taken a "planetary" approach to it. This involves looking at the nature of the interactions between chemistry and biology and the evolution of planets as a whole. This approach should be complementary to those taken elsewhere and, especially, with the vast literature at all levels on the origin of life on Earth.

The book divides naturally up into several topics – the origin and early evolution of life on Earth; environmental conditions necessary for life; the possibility of life on other planets in the solar system; the occurrence of planets around other stars, their habitability, and the possibility of life there; and the nature of intelligent life and the philosophical implications of finding life. Some of these topics are treated in much more depth than others, consistent with our current level of understanding and with the "planetary" approach taken here.

I have not tried to reference each idea back to its original source or to use embedded references. The range of topics is so broad that, if I did, the reference list would comprise several thousand references. Rather, I've provided a representative list of references for each chapter. Some will provide introductory information for someone not familiar with the field at all, while others include some of the most important papers at the most detailed level of understanding.

I would like to take this opportunity to thank those who have helped me to understand some of the issues raised here, sometimes explaining them to me quite patiently. In particular, I very much appreciate discussions with Tom Ayres, Sue Barnes, Mike Carr, Phil Christensen, Bill Cochran, David Des Marais, Jack Farmer, David Grinspoon, Kevin Hutchins, Chuck Klein, Paul Lucey, Jane Luu, Geoff Marcy, Hap McSween, Mike Meyer, Mark Miesch, Ken Nealson, Norman Pace, Cora Randall, Nick Schneider, Everett Shock, Bill Schopf, Steve Squyres, Glen Stewart, and Richard Zurek. I am grateful for comments on early versions of individual chapters or of the entire text to Mark Bullock, Mike Carr, Frank Crary, Kevin Hutchins, Jim Kasting, Geoff Marcy, Chris McKay, Ken Nealson, Frank Palluconi, Nick Schneider, and Len Tyler. Of course, any remaining misunderstandings, inconsistencies, or mistakes are my own. I also would like to thank Adam Black, my editor at Cambridge, for his valuable support and suggestions at all stages of writing this manuscript, and David Underwood in the Graphics Department at CU for his assistance in putting together the images and figures. In addition, I thank Heather Weisacosky for her valuable assistance at all stages of writing the manuscript, including tracking down references and papers in the literature, helping with figures, and reading and editing the entire text.

Last, but certainly not least, I am grateful to Jane Karyl for her support and encouragement during the entire writing process.

Bruce Jakosky

ix

The possible existence of life elsewhere in our solar system or in the universe is one of the most profound issues that we can contemplate. Whatever the answer turns out to be – whether life is present elsewhere or absent – our view of the world and our place within it will be dramatically affected. If a single example of even the most rudimentary life form can be found on another world, that will tell us that the origin of life is not unique to the Earth. We will not be able to help but wonder if intelligent life exists that is more advanced or more sophisticated than us. We will wonder whether the paths that we have chosen as a society are the most beneficial paths. Indeed, it would be a very humbling experience to realize that we are not alone in our existence. On the other hand, if no evidence is found suggesting other life, we may view the Earth and all its inhabitants with a narrow, pinhole focus that makes us unique, on an isolated plane within a vast universe. This view is well expressed in the thought, attributed variously to either the philosopher Bertrand Russell or the science-fiction writer Isaac Asimov, 'There are two possibilities. Maybe we're alone. Maybe we're not. Both are equally frightening.'.

Ours is the first generation that can begin to address in a thorough manner the scientific details of the origin of life on the Earth and the possibilities of the existence of life elsewhere in the universe. Although we do not understand the specific details of the origin of life on Earth, we have learned a great deal about the conditions that surrounded the earliest life. In addition, we have just begun to explore the rest of our solar system and to look for planets outside of it. Most of the planets in our solar system have been visited with spacecraft, and humans have walked on the surface of the Moon. Many details regarding the climate and habitability of the other planets are now understood, and the search for evidence of other life can begin. The *Viking* spacecraft was used to look for life on Mars in the late 1970s – none was found – but we now know that *Viking* might have been searching in the wrong place or with the wrong instruments.

The discovery in 1995 and 1996 of planets orbiting around other stars lends immediacy to the question of life in other solar systems. Although these planets were expected to exist, such expectations were built on theory rather than observation. Their actual detection is a discovery of paramount importance. Even though the new planets are for the most part more Jupiter-like than Earth-like, we can use their existence to begin constraining theories that apply to all planets. Certainly, Earth-like planets – planets with a solid surface, an atmosphere, and the possibility of liquid water on the surface – are expected to exist around other stars, but our telescopes are not yet able to detect them. Knowing that there are Jupiter-like planets, however, gives us impetus to search for Earth-like planets (Figure 1.1).

[Figure 1.1]
Satellite image of the whole Earth. Africa can be seen in the middle of the picture, and Saudi Arabia in the upper right. (NASA Photo.)

Recent advances in our understanding of terrestrial biochemistry also fuel the search for extraterrestrial life. They demonstrate that life can exist within a much wider range of environmental conditions and can utilize a much wider variety of sources of energy than was previously thought (Figure 1.2). They also suggest that there are many distinct ecological niches on the Earth that could have served as the location for the origin of life. Although we do not know for certain where life on Earth originated, we can appreciate that other planets might have life that began in any of several places. If life is as flexible as it appears to be based on our experience on Earth, then it is possible that a wide variety of planets, both within and outside our own solar system, might have originated life and be harboring it at the present. As much as the advances in astronomy and planetary science have opened up our view of the solar system and the universe, the advances in biology have opened up the possibility of abundant life in the universe.

The goal of this book is to present a broad-based view of the scientific underpinnings of the search for extraterrestrial life. And, we know more today than we have at any time in the past about what qualities a planet must possess in order to potentially contain life. We should recognize up front that there is no unambiguous evidence today that suggests the existence of any life in the universe other than on our own planet. This does not mean that there is no life, only that we have just begun the search. Our tools for scientific exploration are just reaching the point where we can

[Figure 1.2]
Photograph of tube worms located near a hydrothermal vent at an oceanic spreading center. They are typically up to two feet long and get their energy, ultimately, from geochemical energy contained in the mineral-rich hot water. (Photo courtesy of J. Childress.)

begin to address these questions meaningfully. The lack of evidence for life should be perceived as an absence of information rather than an absence of life or an indication that life might be unlikely to exist elsewhere.

The field of exobiology (or bioastronomy) – which deals with the study of extraterrestrial life – should be viewed in the broadest sense. Exobiology does not refer only to the study of biological activity on other planets (especially since we know of none today). It also refers to the occurrence of nonbiological and prebiological chemical processes, the distribution of planets within the universe, and the habitability of all planets. As such, it lies at the intersection of the traditional fields of geology, astronomy, plan-

4 etary science, chemistry, and biology. Exobiology is an interdisciplinary field, touching on all aspects of science. And, because of the implications of the results of the search for extraterrestrial life, it also touches on issues that usually are the concerns of philosophy, theology, and other areas that normally are not considered to be a part of natural science.

The idea that life might exist on other worlds is not new to our generation, of course. As far back as the time of the ancient Greeks, several thousand years ago, the possibility of the existence of other worlds created substantial debate. Strong arguments were advanced on both sides. In favor, it was felt that the Earth and life on Earth were simply the agglomeration of a large number of smaller elements, and that these elements could come together elsewhere just as easily. Against, the Earth was felt to be at a special place in the universe – the center – and terrestrial life must be unique. Copernicus was able to argue more than ten centuries later, in the sixteenth century, that the Earth was not the center of the universe. With the recognition that the stars were objects similar to our own Sun, it was felt by many that planets must be located around them – otherwise the stars' existence would be wasted. As recently as the beginning of this century, the presence of advanced life on Mars was generally accepted by the public. Orson Welles' radio broadcast of *The War of the Worlds* in 1936 highlighted this belief and fueled the idea of alien invasion that now has become so popular.

Today, the views on other life in the universe are incredibly diverse. At one end of the spectrum are those who believe that the universe is widely populated by intelligent beings, that the Earth is regularly visited by aliens from outer space, that aliens show themselves to the world in the form of UFOs but have not formally announced their existence, and that humans and other animals are routinely being abducted by aliens for experimental study. At the opposite end, some argue that the Earth is absolutely unique in having life, and that there can be no other occurrence of life anywhere else in the universe. In between, a large fraction of the people – including many, but certainly not all, scientists – believe that life on Earth is a natural consequence of physical and chemical processes, and that life might have arisen independently any number of times throughout the universe.

If there is life on other worlds, the consensus among scientists is that the most likely form to be encountered would be analogous to terrestrial bacteria. On the Earth, bacteria-like, single-celled organisms formed very quickly after the Earth's formation, and they dominated the biosphere literally for billions of years. In fact, it wasn't until about two billion years after the formation of life on Earth, some 2.5 billion years ago, that any

[Figure 1.3]
Hubble Space Telescope image of Mars. The various bright and dark regions are visible, and the north polar ice cap is at the top. (NASA photo courtesy of S. Lee and P. James.)

entities larger than the simplest single cells left their mark; and, it wasn't until about 600 million years ago – after about 6/7 of the Earth's history to date had passed – that substantially more-complex life forms began to appear. In many ways, bacteria still can be considered the most widespread form of life on the Earth. Much of the discussion in this book will focus on the ability of the simplest life to occur and to exist. It is not likely that we will encounter intelligent or advanced beings in our search of the universe, given the low probability of intelligence arising elsewhere. However, this does not negate the importance of finding other life: from the philosophical perspective, finding life of any sort is just as important as finding intelligent life. It would demonstrate that life was widespread within the universe rather than concentrated only on a single world.

The most likely place within our own solar system for finding other life is the planet Mars (Figure 1.3). Although Mars is relatively cold and dry on its surface today, it certainly was not so in the past. The geological evidence that we see shows that liquid water has been present on Mars throughout much of its history. Liquid water is the single environmental requirement thought to be essential for life. With abundant water being present, environmental conditions on early Mars may have been similar to the

6 conditions on the Earth at the same time. This was a time when life was forming on the Earth, and life may have been forming on Mars independently from the Earth. In addition, it is possible that terrestrial life might have been exported to Mars in its early history, by the impact of an asteroid onto the Earth's surface and by the ejection of rocks containing bacteria into space. Even if Mars does not harbor life today, it still is an important target in the search for extraterrestrial life. If there is no life on Mars, and no evidence for a past existence of life, it would then be important to understand what caused the Earth and Mars to differ so dramatically in their outcomes.

There are other possible abodes of life in our solar system, however. It is possible to imagine that life might exist on Europa or Io, both of which are satellites orbiting around Jupiter (Figure 1.4). Although Io does not have any visible evidence of water, it has an obvious source of energy in its abundant volcanic activity. Europa, on the other hand, has large amounts of water – as ice on the surface and maybe as a 'mantle' of liquid only kilometers below the surface. In fact, even Jupiter could conceivably harbor life in its atmosphere, feeding on the organic compounds that are present there (Figure 1.5).

And, of course, there is the possibility of life existing on planets around other stars. Based on theory (and now on observations), we expect to find planets around other stars, and we expect some of them to have conditions that are suitable for the origin and the continued existence of life.

Throughout the rest of this book, we will proceed systematically through the issues pertaining to the possible existence of extraterrestrial life. Of course, such a discussion requires knowledge of the properties of life on the Earth, since that is our only example of life. We will begin with a discussion of the requirements for the occurrence of life on Earth: What were the environmental conditions on the Earth just after its formation? When did the climate and environment first become suitable for the continuous existence of life? What was the earliest life like and, based on the geological evidence, when did it first appear? How does life function? How does it obtain energy from its environment and use this energy to power metabolism and reproduction? In what types of environments can life exist? These will serve as constraints for discussing the origin of life. Although we do not know for certain how life first began, we do have an understanding of what the relevant processes were likely to have been.

Once we understand the conditions that allowed life to occur on the Earth, we can begin to address the question of life elsewhere. Is there life in the rest of our solar system? Rather than proceed through a lengthy discussion of each planet, we will focus on the most likely and most interesting

[Figure 1.4]
Image of the Europa surface, taken from the *Voyager* spacecraft. (NASA photo.)

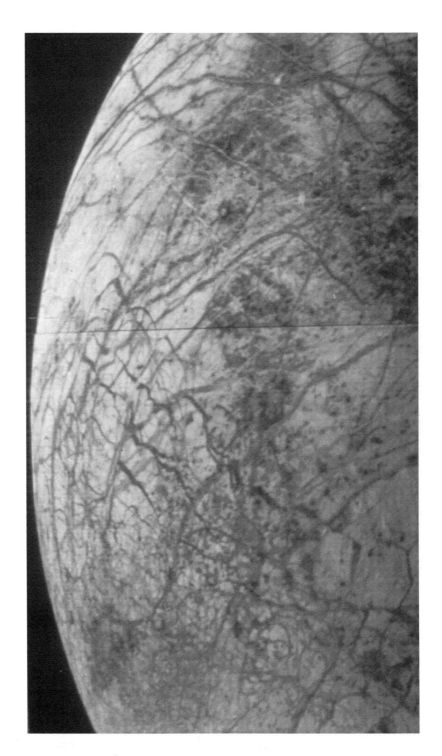

habitats for life. Could life have originated on Mars, and could it possibly have survived up until the present? Can we reach any conclusions about martian life based on the recent analysis of meteorites that are thought to have come from Mars? Although Venus is not a suitable location for life today – because of its extremely high surface temperatures – might it have been at some time in the past? What about Io and Europa, where sources of energy and water might exist and potentially could be tapped by life? What about Titan, a satellite of Saturn, which has an atmosphere rich in organic compounds that may be similar to the prebiotic terrestrial atmosphere? Although not a part of exobiology *per se*, we will include a brief discussion concerning the possibility of 'terraforming' Mars – of changing its environment to one more like that of the Earth, on which terrestrial plants or animals might survive – and we will address the moral and ethical issues involved.

Finally, then, we will turn our attention to the rest of the universe. Based on our understanding of how stars and planets form, do we expect planets, and Earth-like planets in particular, to be a common occurrence? What are the properties of the planets that have been discovered, and what do they tell us about our understanding of the processes by which planets can form? What are the prospects for detecting planets around other stars, and Earth-like planets in particular? What conditions would make other planets habitable, given what we know of the requirements for life? Are habitable planets likely to be abundant, or are they likely to be rare? And, what about the possibility of intelligent life elsewhere in the universe? Is intelligent life likely to exist? How can we communicate with other intelligent species if they do exist?

Life in the context of the universe

In order to begin a discussion of whether life can exist elsewhere in the universe, and to be able to put the relevant issues into the proper broad context, it is useful to begin with a simple framework describing the rest of the universe. This will provide us with a larger context into which we can place the specific details regarding life. In the rest of this chapter, then, we will summarize the origin and evolution of the universe, the galaxy, and the solar system. This will provide us with a sense of the physical scale and of the timescales with which we are dealing.

[Figure 1.5]
Jupiter, as seen from the *Voyager* spacecraft. The Great Red Spot is visible, along with the belts and zones in the clouds. (NASA photo.)

The universe is estimated to have originated about 15 billion years ago. This estimate was calculated by looking at the current rate of expansion of the universe, since all of the galaxies in the universe currently are moving away from each other. Extrapolating backward in time, 15 billion years ago is the time at which the entire universe would have been concentrated in a single point. Because the universe was much denser at that time than it is today, it was much hotter; at time zero, when everything was theoretically at a single point, the temperature would have been infinite. What existed prior to the occurrence of this singularity? There is no way to know. Time would have had very little meaning at that instant, so there may not have been a time prior to the existence of our present universe. Because the universe appears to be expanding from a single point, this theory for the origin of the universe is referred to as the Big Bang theory. Although there still is debate about the exact age of the universe and the rate at which it is expanding, the uncertainties do not affect the overall conclusion that it is expanding or the subsequent history of galaxies, stars, and planets.

As the universe expands, it slows down due to the gravitational pull from all of its mass. Will it eventually slow down enough to stop expanding? If it does, it will then begin to collapse, and would end up the same way it began – as an infinitely dense singularity. We do not know if this will happen – it depends upon the total mass of the universe and the rate of expansion. The mass that we can identify, in the form of galaxies, stars, and clouds of dust and gas, is not sufficient to stop the expansion and make the universe 'closed'. However, there may be large amounts of mass that we have not yet identified – planets that haven't been detected, stellar or substellar objects too faint to be seen, black holes, or even undetectable elementary particles – so no conclusions can be drawn yet as to whether the universe is open or closed.

At the time of the Big Bang, matter composed of atoms as we think of them would not have existed. Rather, the universe would have consisted of pure energy. As it expanded and began to cool off, from initial temperatures so large that the number is essentially meaningless to us, the energy would convert into matter, first to elementary particles and then to protons, neutrons, and electrons. As mass is created out of energy, about 90% of the mass would be in the form of hydrogen and 10% as helium. Essentially none of it would be composed of the heavier elements; these came later. Some of the energy from the earliest periods remained in the form of radiation; as the universe expanded, the radiation cooled and can be detected today as what is called the '3 degree background radiation', which is detected uniformly in all directions in the universe.

As the universe continued to expand, local swirls and eddies of matter produced regions that were dense enough to begin to collapse under their

[Figure 1.6]
Hubble Space Telescope long-exposure image of the sky. All of the objects that are more than points of light are galaxies, with ages ranging from a few billion years to about fifteen billion years, very soon after the universe formed. (NASA photo.)

own gravity. These clouds collapsed and formed the first galaxies and, within them, the first generation of stars (Figure 1.6). The stars were dense enough and hot enough deep within their interiors that nuclear reactions occurred, combining hydrogen atoms together to form helium, and then into heavier elements. These nuclear reactions released energy that was then emitted by the stars in the form of light. As the first stars burned all of their nuclear fuel and reached the end of their lifetimes, in some cases after

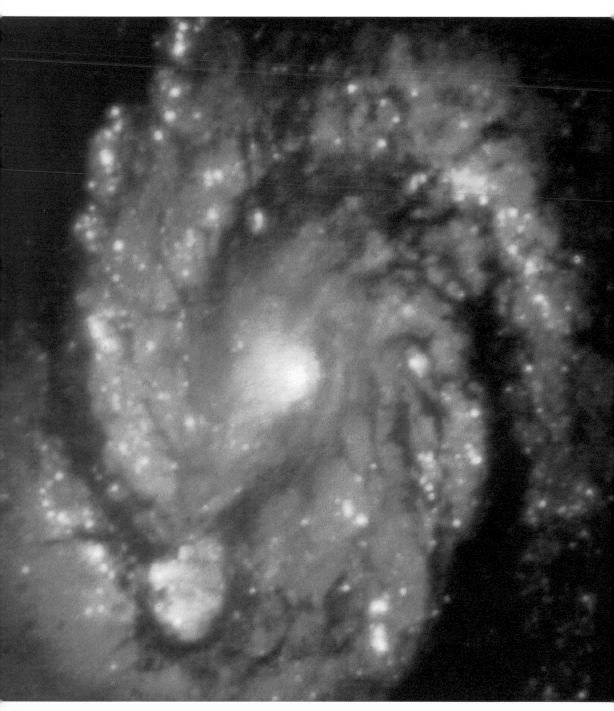

[Figure 1.7] *Hubble Space Telescope* image of the spiral galaxy M100, similar in configuration to our own Milky Way galaxy. (NASA photo.)

only a few hundred million years, they collapsed. Some of them rebounded
from the collapse by exploding cataclysmically, creating supernova explosions. It was in these catastrophic explosions that all of the newly formed
elements heavier than hydrogen and helium were tossed back into space,
only to be incorporated into later generations of stars. The elements out of
which the Earth and all of the living things on it are made were created
within stars; essentially, we are made of recycled stellar material.

This first cycling of material through stars took place quickly. Stars have
been forming ever since, living their lives, and dying either a dramatic or a
quiet death. About ten billion years after the first stars appeared, our Sun
formed from the collapse of a cloud of gas and dust in the outskirts of the
Milky Way galaxy (Figure 1.7), and with it the Earth and planets. This was
about 4.5 billion years ago.

As this cloud of dust and gas around the proto-Sun collapsed, it began
to rotate more rapidly, similar to how a springboard diver or gymnast will
turn faster when pulling into a tucked position. The rotation was sufficiently rapid for some of the dust and gas to be left behind in the collapse,
remaining in orbit around the newly formed Sun. Because of its rotation,
the dust collapsed to form a disk-shaped cloud orbiting around the Sun
(Figure 1.8). The dust was not stable in this disk, however, and it broke
apart into clumps of dust and debris perhaps kilometers in size. These
clumps are called planetesimals (literally, 'little planets'). As they orbited
around the Sun, they began to collide and stick together. As they became
larger and more massive, they were able to attract additional objects by the
force of their gravity, and these 'protoplanets' grew more massive in size
and less abundant in number. They eventually accumulated into a smaller
number of larger objects.

As this accretion process proceeded to completion, the remaining
debris and objects were swept up into what became the planets in our solar
system. The end of accretion saw the last large planetesimals, some up to
perhaps thousands of kilometers in size and worlds in their own right,
collide into the planets with tremendous impact. The entire accretion
process lasted less than a few hundred million years. The main part of the
planets in our solar system formed about 4.5 billion years ago, and most of
the remaining debris was collected by the planets by about 4.0 billion years
ago. Of course, there is still debris in the solar system that continues to
collide with the planets in an ongoing process; the amounts are very small,
however, and contain very little additional mass. (Of course, that does not
mean that an impact of an asteroid onto the Earth today would not be dis-
astrous for its inhabitants!)

The matter responsible for forming the Sun and our solar system was
composed largely of hydrogen and helium with lesser amounts of the

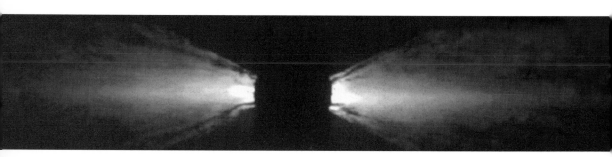

heavier elements. Jupiter and the other giant planets were massive enough to be able to gravitationally attract and hold on to the hydrogen and helium as well as the heavier stuff, and they have a bulk composition similar to that of the Sun. The Earth, however, along with the other rocky planets in the inner solar system, was too light, and very little hydrogen and helium could be grabbed from the disk. As a result, these planets consist primarily of the heavier elements. The Earth is composed largely of oxygen, iron, silicon, and magnesium. There are other important elements, which include hydrogen (which is present but not as abundant as in the Sun), carbon, nitrogen, sulfur, and phosphorous, for example.

Let us put the size of the universe in perspective, starting with the Earth and moving outward. The Earth is roughly spherical in shape, with an average radius of about 6378 km. (By comparison, the continental United States is roughly 4500 km from coast to coast.) The Moon orbits around the Earth, about 384 000 km away; it has a radius of 1735 km and a mass of about 1/81 the mass of the Earth. The Sun is located 150 000 000 kilometers away from the Earth (or, in scientific notation, 1.5×10^8 km; this distance is defined as 1 astronomical unit, A.U.). By comparison, Venus is located about 0.72 times as far from the Sun as the Earth, and Mars about 1.54 times. Jupiter is at a distance of 5.2 A.U. from the Sun, and Saturn is at 9.5 A.U. Pluto, the farthest planet from the Sun, has an average distance of about 39.5 A.U. from the Sun, although its orbit is not quite a circle so that it can be closer or farther at different times.

It takes light from the Sun eight minutes to reach the Earth. The nearest star is more than 250 000 times as far from the Earth as is the Sun; light from this star, alpha Centauri, takes about 4 years to reach the Earth (and, hence, it is four 'light years' away). The galaxy in which the Sun resides is about 100 000 light years across, and the Sun makes one orbit around the center of the galaxy every 200 million years. The galaxy nearest to our own is a few tens of thousands of light years away. Finally, the size of the known universe is about 15 billion light years across. The Earth is actually just a small planet orbiting an ordinary star, in an ordinary galaxy, in a typical corner of the universe.

Life in the universe

We now will begin to explore the issues surrounding the occurrence of life in the universe. Chapters 2 through 6 will address the origin and evolution of life on Earth, and the environmental conditions surrounding the existence of life. This will provide the background necessary to begin a discussion of life elsewhere, and Chapter 7 will discuss the basic requirements that life anywhere would have to meet. Chapters 8 through 13 will discuss the possibility of life existing elsewhere in our solar system, whether on planets or on satellites that orbit around planets. Chapters 14 through 17 will focus on the existence and nature of planets that orbit around other stars, their possible habitability, the possibility of life, and the possibility for intelligent life. The final chapter will include a discussion of the philosophical implications of finding (or not finding) life in the universe, and a brief look at the plans for the future exploration of our solar system and the universe.

The discussion in this book represents the present status of our understanding of the occurrence of life in the universe. Our current understanding of the state of the Earth, the solar system, and the rest of the universe will stand as a benchmark in time. Advances in all aspects of the many different relevant fields are proceeding at a rapid rate, however, and we should expect that new discoveries will continue to be made that will change our view of the universe. We expect that these new discoveries will not make the present discussion obsolete, but will help to fill in the large gaps in our understanding. In addition, though, this book should not be read as a mere collection of facts. Exploring the universe is a process – it is a quest for new horizons expanding one's view of the world. It represents the most visible manifestation of a desire to move beyond the status quo. Therefore, this book should be read as a starting point, not an end in itself. It is the step-off mark for moving outward beyond the Earth, for seeking an understanding of how we as individuals and as a society fit within the framework of the universe. The topics discussed here truly represent the opening up of a new frontier.

16   Before we can begin to look at the processes that are directly responsible for the formation of life on the Earth, we must understand the constraints on the origin of life that are imposed by the early history of the Earth. The Earth formed about 4.55 billion years ago (b.y.a.). Although there is no substantial geological evidence on the Earth that dates back to the first half billion years of its history and can tell us about the environment at that time, we need look only as far as the Moon to see what was happening. The surface of the Moon is geologically old, and few large-scale processes have occurred during the last 3 b.y. that would have disturbed what it looked like during its earliest history. Based on our ability to date rocks brought back from the Moon by the astronauts in the *Apollo* program, we know that the dark 'seas' (termed maria, Latin for sea, or mare in plural form) are the youngest areas. The lunar mare formed between about 3.0 to 3.7 b.y.a. The lunar highlands, on the other hand, generally are older than 3.9 b.y. and record evidence from the earliest history of the lunar surface. For the most part, these highlands are completely covered by craters formed by the impact of debris left over from the formation of the planets (Figure 2.1).

The heavily cratered nature of the oldest regions on the Moon tells us that the lunar surface 4 b.y.a. was constantly being bombarded by asteroid-like objects. Since the Moon was in orbit around the Earth at that time, the Earth must have been undergoing a similar rain of impacting objects. Furthermore, the Earth is more massive than the Moon, and its gravitational pull is more effective in attracting asteroids into it. As a result, the Earth was undergoing substantially more impacts than was the Moon. The impact rate onto the Moon, and presumably onto the Earth, did not decline substantially until between 4.0 and 3.8 b.y.a. Clearly, these impacts onto the Earth would have had a substantial effect on the surface environment, and would have affected the ability of life to exist. The most dramatic climatic effect of these large impacts may have been the heating of the surface of the Earth, the complete evaporation of the world's oceans, and the destruction of any existing life. Thus, the impact history of the earliest Earth provided an environment in which life could not continue to exist even if it had originated.

In this chapter, we will explore the environmental consequences of these large impacts, and discuss the earliest time that life, once formed, could continue to exist without being destroyed. The geological record of the history of the Earth tells us that life existed certainly by 3.5 b.y.a. and possibly by 3.9 b.y.a. This gives us a window in time during which life began, and helps us to understand what the environment was like at that time and how life may have originated.

[Figure 2.1]
The heavily cratered ancient highlands of the Moon, as seen from spacecraft. (NASA photo.)

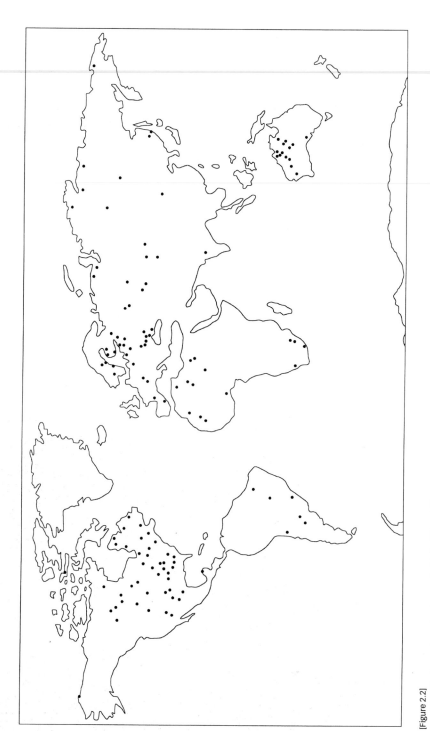

[Figure 2.2]
Location of the known impact craters on the Earth. Although there are
no known craters on the ocean floor, we expect that there must be more
there than on the continents. (After Grieve and Shoemaker, 1994.)

Impacts and the extinction of the dinosaurs

That impacts might affect life on Earth was recognized only recently: the extinction of the dinosaurs and of a large number of other species that existed 65 million years ago may have been caused by the impact of a 10-km asteroid. Because this event occurred relatively recently, from a geological perspective, we can use it as an example of how an impact can affect the environment. We wish to understand what the environmental effects of an impact by various sizes of objects would be, and how often destructive impacts might have occurred in the past (and might continue to occur at the present).

There are about 200 impact craters known on the surface of the Earth (Figure 2.2). These range in diameter from about 1 km to over 140 km. At the small end of the distribution, Meteor Crater in Arizona is just over 1 km across; it was formed about 50 000 years ago (Figure 2.3). Both the Vredefort crater in South Africa and the Sudbury Crater in Canada are about 140 km across; coincidentally, both are almost 2 b.y. old.

Although it seems obvious today that an event that is capable of creating such a large hole in the ground would affect the climate and life, possibly significantly, it took specific evidence to suggest a connection. In the late 1970s, several research groups were trying to determine the cause of

[Figure 2.3]
Aerial view of Meteor Crater, Arizona. The crater is about 1.1 km across and 167 m deep as measured from the rim. It was formed about 50 000 years ago by the impact of a 75-m iron asteroid. Notice the hummocky debris surrounding the crater that was ejected by the impact. (Photo courtesy of D. Roddy.)

[Figure 2.4]
The K/T boundary, seen in sediments deposited about 65 million years ago near Raton in southern Colorado. The clay that constitutes the boundary is the 1-cm-thick layer just below the dark coal seam; the topmost rivet on the knife is centered on this layer.

the boundary between the Cretaceous and the Tertiary geologic periods, about 65 m.y.a. (see Chapter 3 for a brief discussion of the geologic history of the Earth). At this boundary, called the K/T boundary (where T represents Tertiary and K represents the German spelling of Cretaceous), a sudden change in the fossils that are found in the rocks means that a substantial fraction of all living species on Earth died out. Included in this extinction event were the dinosaurs (the most famous of the extinctions), most land-dwelling animals, and a significant fraction of the marine invertebrate species. In fact, this sudden extinction event defines the boundary between the two geologic periods. In most places where the K/T boundary is seen in the geologic record (that is, where erosion has not destroyed it), it is marked by the presence of a layer of clay that can be up to several meters thick (Figure 2.4).

The scientists measured the abundance of what are called the 'siderophile' elements within the clay. Siderophile elements are of interest because they are not very abundant in the Earth's crust – they tended to remain with the iron and nickel when they segregated into the core early in the Earth's history. However, they are abundant in meteorites which never differentiated into a core, mantle, and crust. One of the siderophiles is the element iridium. Within the clay layer, the iridium abundance was found to be ten times its abundance in adjacent layers of rock. This was such an enormous increase that the scientists postulated that a large influx of extraterrestrial material must have occurred at that time. In addition to the iridium and other siderophile elements, they noted that the boundary-layer clay contained grains of the mineral quartz that had been shocked by high pressure and temperature into a denser form. The dense forms of quartz, called coesite and stishovite, occur almost exclusively as a result of the tremendous forces that result from the impact of an asteroid onto the Earth's surface; they are found in abundance at Meteor Crater, for example.

The iridium excess and the presence of shocked quartz suggested that an impact of a single large asteroid onto the surface of the Earth had occurred at the time of the K/T boundary. Based on the worldwide distribution of the clay layer, and on the abundance of iridium within it, it was suggested that the impact of a single 10-km-sized object could have been responsible. A 10-km object impacting onto the surface of the Earth would produce a crater that was almost 100 km across. Unfortunately, there was no known 100-km crater that had formed 65 million years ago. The only known crater of that age was the Manson impact crater in Iowa; this crater is only 10 km across, however, and is much too small to account for the observed amount of iridium.

There were no guarantees that a search for the crater would be successful. The crater, if it existed, might have been obliterated by erosion; and there was about a 40% chance that it had formed on the ocean floor and subsequently had been subducted beneath the continents during the 65 million years since its formation. However, in a remarkable confirmation of the hypothesis, the crater was found, located just off of the Yucatan peninsula in Central America. Known as Chicxulub (pronounced Chicks'-you-lub), it is about 200 km in diameter and lies half on land and half on the ocean floor (Figure 2.5). Lying half on the ocean floor, and having been weathered substantially where it lies above the surface, it had not been identified prior to the search for the K/T impact crater. There is a suggestion in the gravity and topography data that the crater might actually have been about 300 km across. This large a crater would have been formed by the impact of an asteroid 10–20 km across.

What would the effects of such an impact on the climate have been? The minimum velocity that an object colliding with the Earth can have is 11.2 km/s; this is the velocity that would result simply from the free fall of an object through the Earth's gravity field from an infinite distance away. However, because the object had some velocity due to its own motion in orbit around the Sun, the impact velocity would have been larger. A more typical velocity with which an asteroid would hit the Earth would be in the range of 15–30 km/s. For comparison, the muzzle velocity of a rifle bullet is about 1 km/s. Clearly, large impacts contain tremendous amounts of energy, and this energy is released during the collision.

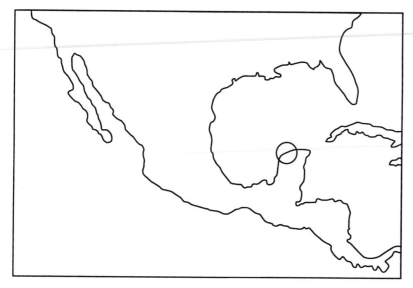

[Figure 2.5]
Map showing the location of the Chicxulub impact crater on the Yucatan Peninsula in Central America. The circle shows the location and size of the ring that is 200 km across; there is evidence for a 350-km ring.

During a typical collision, the energy from the impacting asteroid's velocity is converted into heat and into kinetic energy of motion of debris. A large cavity is carved out as both the asteroid and the planet surface get squeezed aside during the collision. Some of the rock debris can be heated enough to melt or even to vaporize. The debris tossed out during the impact is called 'ejecta'. Ejecta typically falls relatively close to the impact site, within a distance equal to only a few times the diameter of the resulting impact crater. For a large impact, such as the one that formed the 200-km-diameter Chicxulub crater, the ejecta will be thrown on a trajectory that can take it completely out of the atmosphere into space, and it will then fall back through the atmosphere and onto the surface. An impact into the ocean also would result in large-scale mixing throughout the oceans caused by giant tsunamis. In addition, direct heating of the ocean by the impacting object would produce some evaporation of water into the atmosphere.

Clearly, an impact would be devastating to any organisms that lived right where the impact occurred. The heating that occurs during an impact, even if only to relatively modest temperatures, can be sufficient to break apart the chemical bonds within an organism and thereby disintegrate it. In addition, the ejecta that lands near an impact will bury any surface organisms, possibly beneath as much as several kilometers of debris. Dust that is lifted into the atmosphere will absorb sunlight and dramatically change the temperatures both in the atmosphere and on the ground. And, the effects of the tsunamis and the fallout of atmospheric dust will foul any surface or near-surface water that was previously accessible to organisms.

The largest effect on the climate, however, may result from the debris that is ejected into space and then quickly reenters the atmosphere. The larger chunks of debris (kilometers in size, for example) might pass unimpeded back through the atmosphere and create their own impact craters; these 'secondaries' are seen on the Moon surrounding some of the large impact craters. Smaller debris, however, all the way down to dust-sized particles, will be slowed down and heated up on reentering the atmosphere. The dust grains will be heated enough to completely vaporize them. We can see this vaporization of dust grains occurring today as interplanetary dust and meteoroids hit the Earth's atmosphere. We see the incandescent glow of the debris as meteors in the night sky, produced by the friction caused by their passage through the upper atmosphere. For debris from a large impact, however, there would be more than just a few meteor trails in the sky. Rather, the reentering debris would be in the form of a thick blanket of glowing-hot material entering the atmosphere. If one were on the ground watching this ejecta reentering the atmosphere, the entire sky would be filled completely with glowing, incandescent debris. The heat from the sky would be tremendous, and would scorch much of the surface of the Earth. For an impact of the size of the K/T event, the heat could have triggered global fires that would have destroyed many living organisms. This hypothesis is consistent with the discovery of a global layer of 'soot' within the boundary-layer clay.

Dust that was injected into the atmosphere by the impact would have the same effect as a so-called 'nuclear winter'. It would have absorbed sunlight in the atmosphere and kept it from reaching the surface. As a result, the surface would have cooled dramatically in the weeks following an impact, and the temperature may have dropped below freezing for up to a year or more (Figure 2.6). This would destroy the bottom end of the food chain – the photosynthesizing plants – and trigger the death of creatures dependent on them, and, in turn, the death of creatures feeding on the plant-eaters. Over a period of perhaps a year or two, the dust would have dropped out of the Earth's atmosphere and conditions would have returned to normal, but the damage would have been done.

Enough water might have been thrown into the atmosphere by the impact, or evaporated from the oceans by the heat from the ejecta, to produce torrential rains. Although we do not know exactly what effect this would have had, it certainly would have affected all of the land-living species in a dramatic way.

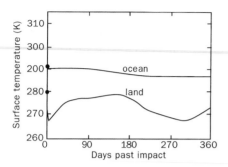

[Figure 2.6]
Average surface temperature on land and the oceans for a year following a large impact. The dots represent the temperatures just before the impact. Over land, temperatures quickly drop below freezing and then rebound before dropping to freezing again. (After Covey *et al.*, 1994.)

The atmospheric gases that were heated up by the impact and by the heat from the impact ejecta would have recombined into different molecules. In particular, the formation of abundant nitrous- and nitric-acid compounds might have occurred. These would have produced a global acid rain that would have destroyed green plants and acidified the oceans.

Conceivably, the impact might have been able to affect the surface of the Earth globally, triggering massive outpouring of volcanic lava. This lava would have released gases into the atmosphere, such as carbon dioxide, which also would have affected the climate. Interestingly, one of the most massive outpourings of lava ever is the Deccan Traps flood basalts in India; these also occurred 65 million years ago, although the large distance between Chicxulub and the Deccan Traps might seem to suggest that the timing was coincidental.

Although the case for an impact-driven extinction of the dinosaurs at the K/T boundary seems strong, there are some nagging problems. The major issue is that the geologic record suggests that the extinctions may not have occurred both instantaneously and simultaneously, and that many of the species may have been dying out gradually even before the impact. Unfortunately, the geologic evidence is not detailed enough to provide a thorough and complete record of the events surrounding this period. It is clear, however, that a large impact occurred at that time, and that the climate effects predicted for such an impact would have been of global proportion and would have had a significant effect on the biosphere. Conceivably, the impact might have been the final blow, attacking a biological system that had already been weakened by environmental stress caused by a different mechanism.

It is also clear that impacts may not have been responsible for all of the mass extinction events in the Earth's history. For example, the extinction that occurred at the end of the Permian period, about 250 million years ago, wiped out about 95% of all living species on the Earth. Recent evidence suggests that the extinction may have been caused by massive outpourings of volcanic material in Siberia. Explosive volcanism would have put gases and dust into the atmosphere that would have affected the climate and possibly triggered the extinctions. Of course, it is possible that an impact occurred at this same time; although no impact crater dates to this time period, it is possible that little evidence would be left behind after such a long time, especially if an impact occurred on the ocean floor.

The rate of impacts onto the Earth

How often will an impact that can affect the Earth's climate occur? The distribution of the sizes of objects in the solar system that are capable of hitting the Earth is described closely by a power law, which means that there are many more smaller objects than bigger objects. For every asteroid that is 100 km in size, there are about 100 objects that are 10 km in size and 10 000 objects that are 1 km in size; in other words, a factor of 10 decrease in size corresponds to a factor of 100 increase in number (Figure 2.7).

The objects that are likely to hit the Earth are those that currently are in orbits that cross the orbit of the Earth. It is only a matter of time, then, until they collide. Today, there are approximately 1000–2000 objects in Earth-crossing or nearly-Earth-crossing orbits that are 1 km in size or larger. Not all of these objects have been discovered yet, but the number of objects can be estimated by using the number of known objects and the rate at which new objects are being discovered . Roughly speaking, about half of the 10-km objects and about 10% of the 1-km objects have been discovered. (Estimates suggest that it would take a devoted effort by a series of a half-

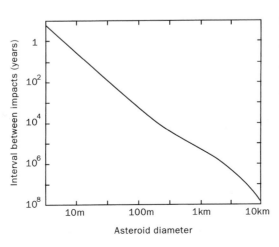

[Figure 2.7]
Power-law distribution of the abundances of objects that are capable of striking the Earth. The figure shows the average interval between impacts as a function of the size of the asteroid. For example, a 10-m or larger object is expected to impact about once every thirty years, and a 10-km object nearly every 100 million years. (After Morrison et al., 1994.)

dozen dedicated observatories working for several decades to discover all of the 1-km-sized objects.)

Based on random statistics, 1-km-sized objects should collide with the Earth on average every 300 000 years. This means that 10-km-sized objects should collide, on average, every 30 000 000 years. Because the asteroid that hit the Earth at the time of the K/T boundary was 10–20 km across, such an impact should occur, on average, every 30–120 million years (Figure 2.7). This seems consistent with the occurrence of the K/T impact event 65 million years ago. Problems would arise in our logic if this large an impact occurred, for example, every 1 million years; we do not see evidence for that many large impacts in the geologic record.

The largest known object that is currently in a nearly-Earth-crossing orbit is several tens of kilometers across. The average lifetime for such an object before colliding with the Earth is around 100 million years. A collision that would be devastating for life as we know on Earth may not occur in our lifetimes, but such a collision probably will occur in the very near future (geologically speaking).

These same statistical arguments suggest that a 100-m object should impact the Earth, on average, every 3000 years, and that a 10-m object should impact every 30 years. The impact of a small comet or asteroid in 1908 in Tunguska in Siberia serves notice to us that these smaller events can occur. Impacts of fist- or boulder-sized objects and of smaller dust-sized objects occur every day.

With a 10-km-sized object hitting the Earth, on average, every 30 million years, why aren't there more impact-caused extinction events? The geologic record of life on Earth provides enough information to search for mass extinction events only since about 580 million years ago (see Chapters 3 and 4). Thus, we might expect to find between 10 and 20 major extinction events during this period that were caused by impacts. In fact, there have been about a dozen mass extinction events during this time period, with five being of truly major significance. Unfortunately, there is not enough evidence to determine the role of impacts in each of these events. Statistically, however, we expect that several of them might have been related to asteroid impacts.

Impacts and the earliest history of the Earth

Now let's turn to the nature of the impact events that were occurring during the earliest history of the Earth. The Earth is thought to have formed by the continual accumulation or accretion of smaller objects. This accretion was caused by the impact of object after object onto the surface until the growing planet became full-sized and most of the debris in space was accumulated onto one of the other planets. (This planet-formation process will be discussed in more detail in Chapters 14 and 15.) The bulk of the accumulation of the Earth was complete by approximately 4.5 b.y.a., and the rate of impact has been declining ever since. These early events are not recorded on the Earth, however, because erosion by wind, rain, and other geological processes has removed all of the evidence.

Because the Moon records impacts from the earliest history of the solar system, without their having been eroded away, we can turn to the Moon as an indicator of the number and timing of the impacts. The largest easily recognized impact crater on the Moon is the Orientale basin, with a rim some 900 km across. There are larger impact basins, however. The largest may be the South-Pole/Aitken Basin, first identified from images and later confirmed from measurements of the lunar topography. It is about 2200 km across and formed nearly 4.1 b.y.a. Remnants of a larger basin may exist as well: there is some geological evidence indicating that a giant impact occurred on the front side of the Moon, encompassing all of the regions that contain the front-side lunar mare. This so-called 'Procellarum Basin' would have been close to 3200 km across and would be one of the oldest features on the Moon (Figure 2.8).

During its earliest history, every new impact crater on the Moon would have covered up some of the existing craters, and only the most-recent events would be observable. In essence, the surface becomes 'saturated' with impact craters. For this reason, we cannot use the lunar surface to obtain information farther back in time than a certain point – older areas would have been destroyed. The oldest areas on the lunar surface date back to 4.4 b.y.a., but a more typical highland age would be close to 4.1 b.y., and the youngest highland surface would be about 3.9 b.y. old. By 3.5 b.y.a., the impact rate had dropped to approximately one per cent of that at the time of the formation of most of the highlands. Only occasional impact events have occurred since then (Figure 2.9).

The impact rate on the Earth is greater than that on the Moon, because of the larger size of the Earth and its greater gravitational pull. Roughly speaking, about two dozen times as many impacts would have occurred on the early Earth as on the Moon. This means that, if the South-Pole/Aitken Basin was the result of the largest impact event on the Moon subsequent to about 4.4 b.y.a., then it is likely that there would have been about two dozen impacts of similar size onto the Earth. To put this in perspective, the creation of such a basin on the Moon would have required the collision of

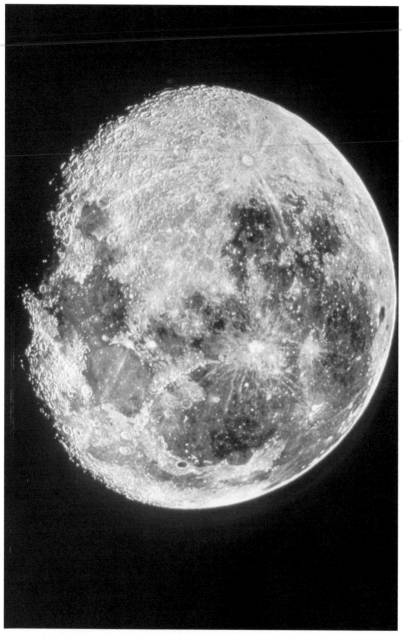

[Figure 2.8]
The nearly full Moon, as seen from Earth. The occurrence of all of the mare regions on one side of the Moon may be explained by the presence of a large impact basin (the 'Procellarum' basin) on the front side. This impact would have thinned the crust substantially, and later impacts would have provided additional thinning, thereby allowing magma to work its way easily up to the surface.

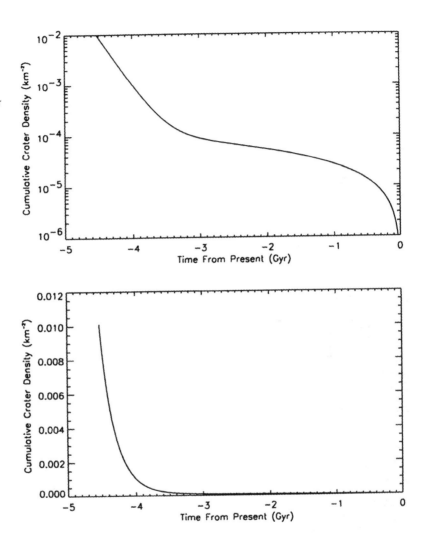

[Figure 2.9]
The number of impact craters that would be present on the Moon on surfaces of different ages. The top panel shows the number on a logarithmic graph, and the bottom shows the same number on a linear scale. The actual impact rate onto the Moon would be the slope on this graph, and would have been largest early in history and smaller more recently. (After Melosh and Vickery, 1989.)

an object approximately 200 km across; this is bigger than all but about a dozen of the asteroids that currently reside in the asteroid belt that orbits the Sun between Mars and Jupiter. And, of course, these large impacts would have been accompanied by a much larger number of impacts by smaller objects.

These colliding objects would have been much larger than those that were responsible for the extinction event at the K/T boundary, up to thousands of kilometers across. Although the means of affecting the climate would have been similar to those of the smaller impacts, the environmental effects of these larger impacts would have been even more dramatic. They would have had a tremendous effect on any life that might have existed during the earliest periods of the Earth's history.

The most significant effect on the climate may have been from the heat deposited into the atmosphere during the impact. The heat would have come both from the impacting object itself and from the reentry into the atmosphere of impact ejecta. The atmosphere would eventually radiate all of this excess heat energy to space and cool off; at the same time, it would radiate energy to the Earth's surface. This surface heating would warm up the ground and ocean surface, and would cause large amounts of water to evaporate into the atmosphere.

Computer simulations suggest that nearly half of the energy involved in the collision would be buried as heat in the vicinity of the impact itself, half of the remaining energy would heat up and evaporate the ocean surface, and the remaining energy would be radiated to space. This allows us to estimate how much energy is used in heating the ocean, and how much water would evaporate into the atmosphere. The impact of an object about 350–400 km across would eject enough debris to provide sufficient heat to evaporate all of the world's oceans and to heat the surface of the Earth to a temperature of more than 2000 K. Of course, if all of the water from the oceans was in the atmosphere as water vapor, the gas would act like a greenhouse and would trap the heat from the Sun. The atmosphere would remain hot, and the water vapor would take about 3000 years to cool and condense back into an ocean. Clearly, such an event would have devastating consequences for any biological organisms inhabiting the surface.

A smaller impact might not vaporize the entire ocean, but could evaporate the water in the uppermost 200 m of the oceans. This depth is referred to as the photic zone; it is the depth to which sunlight can penetrate and drive photosynthesis. If the earliest life were photosynthetic, it would have to occupy this region of the oceans, and the evaporation of the photic zone in the oceans would cause their destruction. As these organisms would be at the bottom of the food chain if any more-advanced organisms had formed, this would have equally devastating results for all life. It would take the impact of an object 150 to 190 km across to supply enough heat to evaporate the photic zone.

Recall that some two dozen impacts sufficient to evaporate the photic zone in the oceans would have occurred since 4.4 b.y.a., and that a few larger impacts would have occurred. Based on the size distribution of the craters on the Moon, it is possible that the largest object hitting the Earth since 4.4 b.y.a. was between 1000 and 1700 km across. This should not be too surprising because the Earth is thought to have accreted from relatively large, lunar-sized objects (see Chapter 14).

To summarize this discussion, it is clear that impacts were occurring during the period up to about 3.8 b.y.a., and that the largest of these impacts would be capable of sterilizing the Earth by evaporating the oceans completely and heating the surface to a temperature of more than 2000 K. When did the Earth first become continuously habitable? This is equivalent to asking when was the last impact that was large enough to either evaporate the oceans or to evaporate the photic zone of the oceans.

Statistically, there probably were about five Earth-sterilizing impact events since 4.4 b.y.a. The last such event could have occurred as long ago as 4.4 b.y.a. or as recently as 3.8 b.y.a. Keep in mind, though, that we are dealing with the statistics of small numbers, and these are inherently uncertain. In this context, it is important to note that there are still four objects in the asteroid belt that are big enough to evaporate the oceans if they were to collide with the Earth. These are the asteroids Ceres, Pallas, Vesta, and Hygiea. Although the asteroid belt may seem quite far removed from the Earth, asteroid orbits are continually being perturbed due to the gravitational influence of Jupiter. The orbits of some of the asteroids, and of debris from the asteroid belt, are capable of being changed over time into orbits that cross the Earth's orbit, making collisions with the Earth possible. Thus, we cannot say for certain when the last ocean-evaporating impact event occurred, and we do not know if one might occur at some distant time in the future.

Because impact events capable of evaporating only the photic zone (upper 200 m) of the oceans are caused by smaller and more numerous objects, we can place more stringent constraints on their occurrence. Again speaking statistically, the last photic-zone-evaporating event probably occurred as recently as about 3.8 b.y.a. And, there still are more than 60 asteroids in the asteroid belt that are larger than 150 km in diameter, and a very large number of smaller ones (Figure 2.10).

These conclusions are not as certain as we would like, however. Alternative calculations allow for the possibility that the number of impact events may have been substantially larger during this time period. The difference between the two calculations is the uncertainty in trying to derive an impact flux onto the Earth based on the number of impacts that are recorded on the Moon. Using a larger but not necessarily unrealistic flux, it is possible that the last ocean-evaporating event occurred as recently as 3.5 b.y.a., and that the last photic-zone-evaporating event occurred substantially later in time. This would complicate the situation because of the geological evidence for the existence of life 3.5 b.y.a. and possibly as long ago as 3.9 b.y.a. (see Chapter 4). If an impact event that was capable of destroying terrestrial life occurred after 3.5 b.y.a., it would

[Figure 2.10]
The asteroid Ida, imaged by the *Galileo* spacecraft en route to Jupiter. Ida is about 60 km long in the longest dimension. (NASA photo.)

require that the biota recorded in the geological record not form a continuous chain of life up to the present. Instead, life would have to have been created and destroyed multiple times during the earliest part of the Earth's history. However, it is important to note that the last major impact could not have been much later than 3.5 b.y.a. without having left some evidence of its occurrence in the geological record.

Although there is much confusion based on the uncertainties in the impact flux and in the geological record, it is clear that the flux of impacting objects onto the Earth's surface does constrain when life could exist. If the earliest life lived in the deep ocean, for example within hydrothermal environments around ocean-floor volcanic centers, then it could have lived continuously since the last ocean-evaporating event. Such an event most likely would have occurred sometime between about 4.4 and 3.8

b.y.a. Organisms that thrive in these conditions are seen today, living deep beneath the surface of the Earth or in extremely hot water circulating around recent volcanic intrusions (hydrothermal systems or hot springs). As will be discussed in Chapter 5, these organisms may represent the oldest type of terrestrial organism.

Alternatively, if the earliest life existed in the photic zone of the oceans, then it must date from the last event that was capable of evaporating the upper surface of the oceans. This last event would have occurred about 3.8 b.y.a., and possibly more recently.

The role of Jupiter and Saturn

Now that we have considered both the giant impacts that occurred in the earliest history of the solar system and the smaller yet still significant impacts that occurred throughout geologic time, let us consider one additional aspect of the impacts: What would the history of impacts onto the Earth have been if Jupiter were not present in our solar system? And what would the effects have been on the Earth's biosphere?

A major effect of the presence of Jupiter and Saturn has been to gravitationally 'eject' comets and asteroids from the solar system. The comets started out in orbits that reached as far as 10^5 A.U. from the Sun (however, see Chapter 14 for a discussion of how the comets came to be at that distance from the Sun). At that distance, their orbits are perturbed by the passage of nearby stars, with some of the comets then entering the inner solar system. When their orbits reach into the main part of the solar system and cross the orbits of Jupiter and Saturn, the comets can be gravitationally perturbed by the large planets. The result most often is the complete ejection of these objects from the solar system. Similarly, a substantial fraction of the asteroids whose orbits are perturbed by Jupiter and Saturn can be ejected from the solar system.

If Jupiter and Saturn were not there, or if in their place there were a larger number of smaller planets that would not have the same gravitational effect, these comets would not be ejected and would enter the inner solar system. There, they would eventually impact onto one of the terrestrial planets. If this were the case, the present-day rate of comet impact onto the Earth could be as much as 10 000 times its actual current rate. Impacts of the size that created the K/T boundary would be expected to occur every 100 000 years instead of every 100 000 000 years. This would have to have a dramatic effect on biological activity – it is not clear if the terrestrial biosphere could withstand a major perturbation that often and still survive. Or, if it did survive, it is possible that only the simplest organisms could thrive. Without Jupiter and Saturn, Earth might be inhabited only by bacteria!

Concluding comments

Clearly, the nature and timing of impacts of asteroids onto the Earth's surface throughout time has played a major role in its history. Early in the Earth's history, the large rate of impacts would have continually destroyed any life on the Earth and kept life from gaining a foothold. Only as the impact rate decreased, about 4.0 to 3.8 b.y.a., would life be able to live continuously without being destroyed. Throughout the remainder of history, however, impacts have continued to play a role. The impact of an asteroid certainly played a major role in the extinctions at the K/T boundary. And, impacts may have played important roles in the dozen or so other mass extinction events that were recorded during the last half-billion years. Looking to the future, it is clear that impacts will continue to occur onto the Earth, and that they will have a tremendous environmental impact (so to speak) when they do.

The harsh environment created by the impact events associated with the origin of the Earth kept life from being able to survive until well after the bulk of the Earth had formed (Chapter 2). Depending on exactly where and how life formed – whether in a shallow-water environment or on the deep ocean floor, for example – impacts may have continually sterilized the Earth until sometime between 4.4 and 3.8 billion years ago. In order to understand how life formed following this violent period, we must understand the nature of the geological processes that were occurring on the Earth at this time. To help us identify the processes that were active before 3.5 b.y.a., we begin by discussing the geological and geophysical processes occurring at the present; then we can discuss how these processes might have operated earlier in history. Once we have this context, we will discuss (in Chapter 4) the history of life as it appears in the Earth's geological record. Together, these will tell us when life formed, what characteristics it would have to have had in its earliest history, and how it might have interacted with its environment. A quick look at some of today's organisms that are able to thrive in extreme environments will give us insight into the types of harsh environments that life is capable of inhabiting and the chemistry that it is capable of employing (Chapter 5). These concepts will then allow us to look at the various processes that may have played a role in the origin of life (Chapter 6). Although we could jump immediately to a discussion of the origin of life, the wide variety of processes and observations that we will deal with first will provide important constraints on the origin.

The dynamic nature of the Earth

The present-day geology of the Earth is dominated by processes that are driven, ultimately, by heat from the Earth's interior. This suite of processes – including the evolution of the Earth's core, the decay of radioactive elements within the interior, the convection of the Earth's mantle to get rid of heat, and the movement of large plates at the surface of the Earth in response to this convection – has played an important role in the nature of the surface, one that has been recognized only within the last few decades.

Geophysical thought underwent a revolution in the 1950s and 1960s as scientists realized that the surface of the Earth – the continents, mountains, and the ocean floors – was dynamic rather than static. The idea that continents could move around the surface of the Earth, even though such movement occurred only on timescales of tens or hundreds of millions of years, represented a true scientific revolution. The evidence supporting

[Figure 3.1]
The puzzle-like fit of North and South America, Africa, Greenland, and Europe, indicating how they used to be a single continent. The outline shown is that of the current sea level on the continents; the fit would be better if the outline at the base of the continents were used. Also, the fit is not exact because of distortion of the continents following the break-up. (After Monroe and Wicander, 1995.)

the movement of continents and the steady creation and destruction of the ocean's floor became overwhelming.

The puzzle-like fit of South America into Africa was just one example of the evidence that continents might have the ability to move (Figure 3.1). This fit suggested that these continents at one time had been part of a larger super-continent, and that this large land mass had broken into pieces that were subsequently pulled apart. By itself, however, this evidence was unable to overcome the prevailing scientific opinion that the rock making up the Earth was too strong to move in response to any conceivable pushing force. Not until later was additional, stronger, geolog-

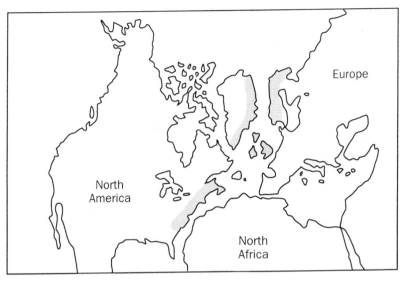

[Figure 3.2]
Diagram showing the continuity of mountain ranges across the continent boundaries when the continents are lined up in their best fit as in Figure 3.1. The shaded areas show mountain regions in the eastern United States, eastern Greenland, Great Britain, and Norway. (After Monroe and Wicander, 1995.)

ical evidence uncovered that suggested that some of the continents had at one time been located together.

In some places, the types of rock that were present on the continents appeared to be continuous across continental boundaries, even though the continents were separated. When the maps of South America and Africa were pushed back together, rock types on one continent matched up against the same rock types on the other. Again, this suggested that the continents had been together at one time and that the intervening ocean floor had not been there. Some mountain ranges also appeared to be continuous across continental boundaries. Similar types of mountains occurred, for example, in the Appalachians (in North America), Greenland, Ireland, Britain, and Norway (Figure 3.2).

There is also geological evidence indicating the presence of ice sheets along the southern tips of South America and Africa, with the direction of movement of ice being toward the land rather than away from it. As this would appear to require that ice move from the oceans up onto the land, opposite to what is observed to happen, an alternative mechanism was required. The suggestion was made that these areas all were once connected to Antarctica, where the ice flows northward, away from the land. If the continents all abutted against each other at one time, then, the ice could flow from Antarctica directly onto the tips of South America and Africa.

In some places, fossils were found of various animals that had a limited

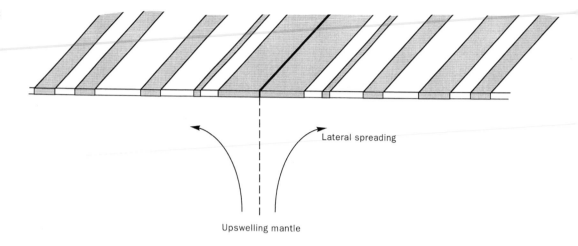

Lateral spreading

Upswelling mantle

geographic distribution, but their occurrence again appeared to be continuous across continental boundaries. All of these instances would be extremely difficult to explain unless the continents had been together at one time and had later moved apart.

The most compelling evidence for movement of the continents, and what finally provided enough evidence to persuade most scientists of this revolutionary idea, was exhibited in the Earth's magnetic field. Volcanic rocks, as they cool and solidify, take on an imprint of the Earth's magnetic field as it exists at that time, and they 'remember' what that magnetic field was. The orientation of the Earth's magnetic field varies over time, causing rocks of different ages to have imprints of a magnetic field with different orientations. By measuring the imprinted magnetic field within rocks of different ages, then, one can reconstruct the history of the orientation of the magnetic field – essentially, the history of the location of 'magnetic north'.

Since all rocks anywhere on Earth experience the same magnetic field, the history determined from rocks in North America, for example, should agree with that determined from rocks in Europe. However, this history appeared to be different when it was derived from rocks on the different continents. Although the magnetic field orientations agreed with each other for the most-recent rocks, the older rocks suggested that the continents had experienced somewhat differing orientations of the magnetic field. These results could be reconciled only if the continents were moving with respect to each other. The older the rocks that were examined, then, the more that the derived magnetic fields would differ.

At this same time, magnetic surveys were being done of the sea floors,

[Figure 3.3]
Schematic diagram showing the remnant magnetic field in the sea-floor crust. The crust is created at ridges above upwelling mantle, where molten rock is injected at the surface; there, it takes on an imprint of the Earth's magnetic field at the time that it crystallizes. The changing polarity of the field (whether magnetic 'north' points toward the geographic north pole or south pole) then can be seen in the crustal rocks. The occurrence of regions of constant magnetic field as stripes, parallel to the ridge, and the occurrence of the same pattern of changes on both sides of the ridge, is best explained by the creation of new crust at the ridge and its spreading laterally away from the ridge.

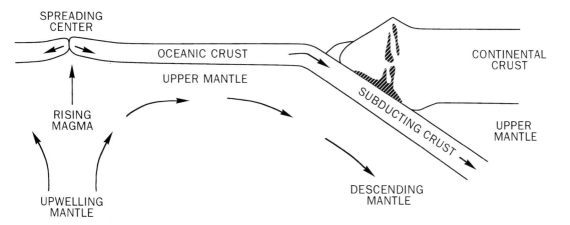

[Figure 3.4]
Schematic diagram showing the spreading of oceanic crust that occurs above upwelling mantle regions and the subducting of oceanic crust beneath continents. Both processes are driven in part by the convective overturning of the mantle. The shaded regions indicate where rock that has been melted in the subduction zones is upwelling to the surface to form volcanoes and contribute to the growth of continents.

and 'stripes' of different magnetic field orientations were discovered. Not only was the direction of the Earth's magnetic field changing (with magnetic north being at the north geographic pole at one time and then changing to the south pole at another), but the seafloor showed this changing orientation in the form of coherent stripes that had the same orientation. In addition, these stripes were 'mirrored' across ridges down the middle of the oceans – the magnetic field showed the exact same pattern on each side of the ridge (Figure 3.3). Also, these stripes corresponded to rocks of different ages. The youngest rocks, which solidified very recently, were located adjacent to the mid-ocean ridges, and progressively older rocks were found farther away from them. It rapidly became clear that the only plausible explanation was that the rocks on the sea floor were being created along these ridges by the injection of new volcanic rock into the crust and that, with time, these new rocks must be moving away from the ridges. These mid-ocean ridges occur in the Atlantic, the Indian, and parts of the Pacific Oceans, and together span a length of over 20 000 km.

Although the suggestion that South America and Africa fit together was made during the early part of the century, it wasn't until the mid-1960s that this new picture of a dynamic surface finally crystallized and was accepted by the scientific community. And, just within the last decade, the movement of the continents has been measured directly using satellite observations. Today we have a much more-detailed perspective on the Earth's surface: crustal rock is being created continuously at mid-ocean ridges by the injection of melted rock from the upper mantle. In order to make room for this new crust, the older crust is moving outward, away from these ridges, at an average rate of about 10 cm/year (Figure 3.4). These ridges are known as spreading centers because of this movement. Since the Earth is

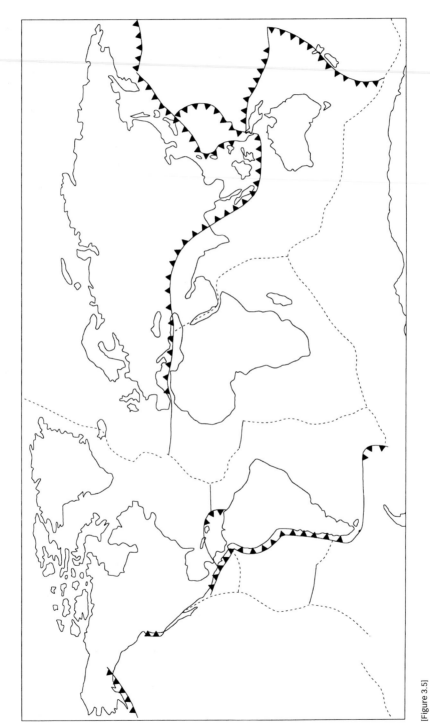

[Figure 3.5]
Global map of the Earth indicating the locations of subduction zones (solid lines with 'teeth', with the teeth pointing in the direction of subduction), spreading centers (dashed lines), and transform faults (solid lines). Continents are shown by their outline.

not growing larger, the creation of crust in some locations must be bal- anced by an equal amount of destruction of crust in other places. Crust on the ocean floor is consumed in some locations where it pushes up against continents. In a process called subduction, the crust is pushed underneath the continents and back into the upper mantle (Figure 3.4). The continents are less dense than either the ocean-floor crust or the upper mantle, so they tend to 'float' above them at the surface. The sea-floor crust, which is denser than the continental crust, easily sinks beneath the surface. The picture is a little more complicated than this when one looks in detail at some of the specific subduction zones, but this general picture serves to describe the basic nature of the process.

A typical example of a mid-ocean spreading center is the mid-Atlantic ridge (Figure 3.5). It extends down the middle of the Atlantic Ocean, and is located about half-way between North and South America on one side and Africa and Europe on the other. New crust is being formed here, and the two sides of the Atlantic are moving steadily away from the ridge. The ridge itself pokes above the ocean's surface in Iceland, and this is the only place on Earth where the creation of this type of crust can be seen directly. The volcanic nature of Iceland results from this spreading process (Figure 3.6).

Subduction zones, where crust is being lost, line most of the western Pacific Ocean and parts of the eastern Pacific (Figure 3.5). The subduction process occurs a little bit at a time, and the sudden, jerky movements of crust as it happens are one cause of earthquakes. For example, movement associated with subduction of the Pacific beneath the eastern edge of Asia is responsible for the devastating earthquakes in Japan. Not all movements of crust involve its creation or destruction, however. The well-known San Andreas fault in California, for example, marks where two plates are sliding past each other (Figure 3.7).

The mid-ocean ridges, subduction zones, and transform faults (like the San Andreas) mark the locations of boundaries between large, coherent plates of crustal material. There are about a dozen such plates on the surface of the Earth, ranging in size from the entire Pacific Ocean basin to only a couple hundred kilometers across for the Juan de Fuca plate that is being subducted beneath the Pacific Northwest region of the United States. The movement of these plates, and the forces that drive them, are referred to collectively as plate tectonics. The word 'tectonics' refers to the movement of material at the surface of the Earth, and 'plate' involves the presence of discrete plates of crustal material.

As they move at the surface, the continents and the ocean floors are not sliding on top of the mantle. Rather, the underlying mantle material is moving along with the plates. This motion results from a steady over-

[Figure 3.6]
Aerial photo of the Icelandic
volcano Vatnajokull. Iceland is
the only region where mid-ocean
spreading-center volcanism occurs
above the ocean surface. (Photo
courtesy of M.T. Guomundsson.)

turning of the mantle. At the high pressures and temperatures that occur within the Earth's interior, rock will deform smoothly rather than break suddenly when it is under stress. In our own experience, silly putty or taffy tend to behave this way – if you pull slowly on silly putty, it will stretch rather than break; only when you jerk suddenly will it snap.

The mantle is overturning in order to remove heat from the interior, in a process termed 'convection'. There are four sources for this heat, although only two are ongoing at the present.

The decay of radioactive elements is an important source of heat for the Earth's interior. The radioactive elements – uranium (abbreviated U), thorium (Th) and potassium (K) – are present in very small amounts as contaminants in most rocks. The nucleus of each atom is unstable and decays over time to another element by the emission of sub-nuclear particles from the nucleus. Each decay also releases a tiny amounts of heat. By itself, the heat released from the decay of a single atom is insignificant. Given the large total amounts of radioactive elements when summed over the entire Earth, however, the total amount of heat generated is substantial. For example, at the current rate of heating, if the heat had not been removed from the Earth's interior through time, the temperature of the Earth would have risen several thousand degrees since its formation. This would be enough heat to have melted the Earth thoroughly. Of course, as the elements decay, there are fewer of them left to decay further; this means that the rate of decay and the rate of heat production were greater in the past, perhaps several times greater than today's heat production rate.

The second heat generation mechanism is the freezing out of the inner core of the Earth. The Earth's core is composed predominantly of a mixture of the metals iron and nickel. The inner part of the core is solid, but the outer part is molten. As the Earth loses heat with time, the molten part of the core is freezing, giving up its heat in the process. Eventually, the molten part of the core will freeze entirely, and this source of heat will no longer be important. It is the freezing of the core, and the overturning of the outer core as a way of getting rid of heat, that is thought to be responsible for the formation of the Earth's magnetic field. Once the entire core becomes solid, there will no longer be a mechanism for generating a magnetic field; this is a long time off, however.

The third source of heat operated only during the formation of the Earth – the heat provided by the impact of all of the objects accreting onto the forming Earth. As described in Chapter 2, each impact would have deposited some of its energy into the planet's surface, thereby heating it up. This source of heat alone would have been sufficient to melt much of the outer portions of the growing planet.

The final source of heat also operated only during the Earth's early history. The Earth probably formed as a homogenous body. Subsequently, the heavier iron and nickel settled to the center of the Earth to form the core. As it did so, potential energy was released as heat, heating up the Earth. This process can be thought of in a simple manner. As the heavier iron settled toward the center of the Earth, it did so by trying to fall through the rest of the rocky material. Friction kept it from moving very fast, and this friction turned the energy from falling into heat. The heat from forming the core also would have been enough to have melted the entire Earth. However, the heat supplied by both core formation and accretion would have dissipated relatively quickly, and very little of it is left today. The only sources of heat at the present are radioactive decay and solidification of the core.

The overall result of all of this heat production is the heating up of the

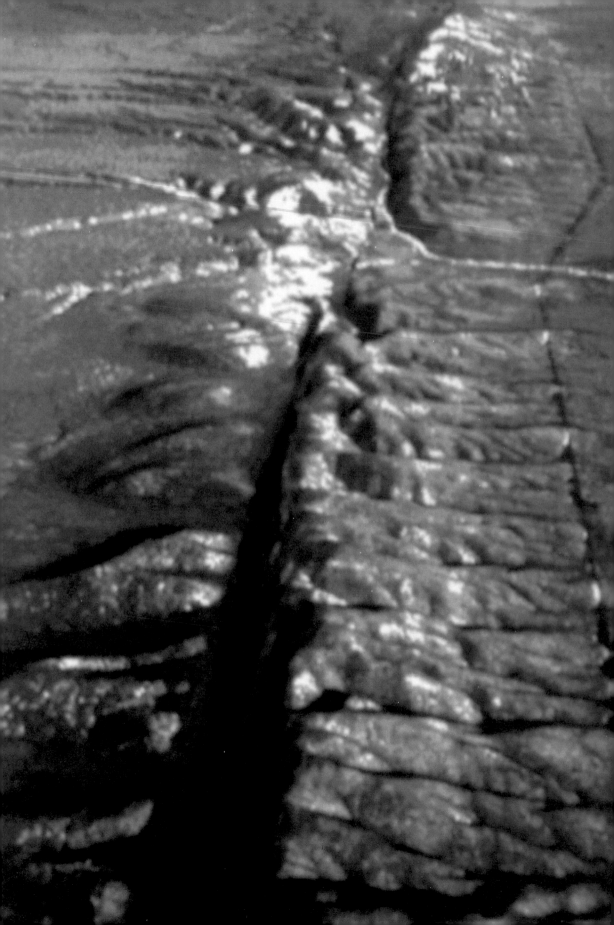

Cooling at the top

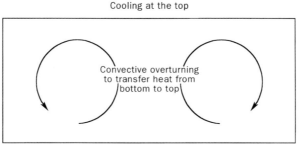

Convective overturning
to transfer heat from
bottom to top

Heating at the base

[Figure 3.8]
Schematic diagram showing how solid-state convection works in the Earth's mantle. Heat is applied at the bottom, creating a temperature contrast with the top. This drives an overturning, in which the cool material at the top is replaced by warmer material from below. The near-surface material then can cool off by conduction of energy to the surface. Continued heating at the base and cooling at the surface maintains the temperature contrast, driving a continuous overturning.

Earth's interior. As the interior is heated, it expands a small amount and becomes less dense. The warmer, less-dense material near the bottom of the mantle wants to rise as the cooler, denser material at the top wants to sink. As a result, there is a general overturning of the mantle, with the rock at the top changing places with the material at the bottom. This overturning does not happen just once, however. Because heat is being supplied continuously, the mantle always will have a tendency to overturn to get rid of the heat. This continuous overturning is the convection of the mantle (Figure 3.8). The convection brings the hot material from deep inside the Earth closer to the surface. Once near the surface, it can cool off quickly by conduction to the surface and then to space, thereby getting rid of the excess heat.

This convection of the mantle expresses itself at the surface as horizontal movements of the crust. The region of the surface that is located between where the mantle is upwelling in one location and downwelling in another contains laterally moving crust. Where upwelling mantle approaches the surface and begins to spread laterally, the crust is pulled apart by this lateral movement. New crust is injected into the surface to fill the gap that results from the pulling apart of the crust; this results in the formation of a spreading center. The new crust that is created is a type of rock that is denser than the continental crust; as a result, it does not stand as high as the continents as it floats on the underlying mantle, the water in the oceans settle into these low areas, and they become the ocean floor. This difference in the height that the rock stands above the surface is similar to the behavior of objects floating on water. An air-filled ball, for example, is less dense than an ice cube, which is itself less dense than liquid water. As a result, a ball floating on water will stick up higher than would an equal-sized block of ice.

Where two crustal plates are being pushed together, something has to give. If denser oceanic crust is being pushed up against less-dense conti-

[Figure 3.7]
Aerial view of the San Andreas Fault in California. West is to the left side of the figure. The Pacific plate on the west is moving northward at an average rate of about 5–10 cm/yr. (USGS photo by Robert Wallace.)

46 nental crust, the former will be pushed beneath the latter. The oceanic crust has to go somewhere, and it goes back into the upper mantle; in part, the crust pulls itself down in to the mantle because it is dense enough to 'sink'. Whether it stays in the upper mantle or ultimately ends up in the lower mantle, accumulating at the core–mantle boundary, is a subject of current debate. This process of pushing the ocean floor back into the mantle is termed 'subduction'. Occasionally, two continents will be pushed up against each other. In this case, neither one is dense enough to be pushed beneath the surface and back into the mantle, so they both end up being 'scrunched' up against each other. This is occurring today where India is in the process of colliding with Asia; the 'scrunching' results in very high mountains being produced – the Himalayas.

 Another aspect of the heating of the interior includes the production of volcanic hot spots at the surface of the Earth. These are locations where narrow plumes of hot rock rise directly from deep within the mantle all the way to the surface. There, they result in a very localized region of volcanism that is termed 'hot-spot volcanism'. Two of the best examples of hot spots on the Earth are the Hawaiian Islands and the Yellowstone region. The Hawaiian Islands are a chain of volcanoes that have been created by the movement of the Pacific plate over a rising hot-spot plume (Figure 3.9). Because the source of the hot plume is deep within the mantle, the surface of the plate can move with respect to the plume. The resulting chain of volcanoes stretches back, primarily as undersea seamounts, to almost as far north as the Aleutian Islands that string out away from Alaska; this is a dis-

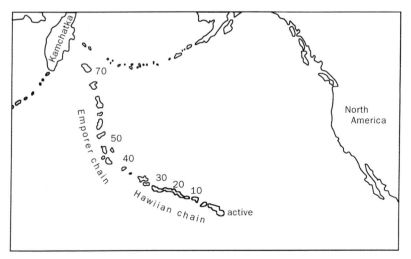

[Figure 3.9]
Map showing the Hawaiian–Emperor chain of islands and seamounts that stretches from the mid-Pacific all the way to Kamchatka. The ages of the rocks in the chain in millions of years is indicated, leading to active volcanic eruptions at the south-easternmost end of the chain. The best explanation for the occurrence of a discrete chain of seamounts and islands, and for the trend in ages leading to the youngest at one end, is the existence of a source of volcanic rock that is located within the mantle and that is moving with respect to the Pacific sea floor. (After Macdonald et al., 1983.)

tance of over 4500 km. Only the south-easternmost islands of the Hawaiian chain are volcanically active today (the island of Hawaii and, to a lesser extent, Maui), although a new undersea volcano off the southeast coast of Hawaii is currently continuing this trend of new mountains.

Yellowstone was produced by hot-spot volcanism associated with the rise of a mantle plume beneath the North American continent. There, the hot spot has moved over the last few tens of millions of years from the state of Washington to Wyoming. Yellowstone is not volcanically active today, but molten magma still exists relatively near the surface; this hot intrusion drives the movement of water through the crust, producing the geysers and hot springs that make the region famous. There is every reason to expect that Yellowstone will continue to be active in the future, both as a geothermal region and with active volcanism, and that it will continue its migration to the northeast.

One of the important aspects of plate tectonics is the role that it plays in producing hot water near the surface of the Earth; as will be discussed in Chapters 5 and 6, this role may have been important in the origin and early evolution of life. At the mid-ocean ridges, water from the oceans can percolate into the crust. There, it is heated by near-surface magma. The hot water rises back up through the crust and is injected back into the oceans. This does not play a role in heating the oceans overall, as the heat is lost quickly to the atmosphere and then to space. However, where the hot water is returned to the ocean, its temperature can be extremely high; temperatures up to 450 °C have been measured. The water does not boil, despite the high temperatures, because the weight of the overlying ocean keeps it liquid. And, at this high temperature, large quantities of minerals can be dissolved in the water; as the water cools off, some of the minerals will solidify and create small cones of rock on the ocean floor, showing where hot water had vented into the ocean. Based on estimates of the number of vents at the mid-ocean ridges and how rapidly water passes through each vent, it takes only one to ten million years for all of the water in the oceans to pass through the vents and be subjected to this heat.

At Yellowstone, hot springs and circulation of water through the crust occur despite the lack of a nearby ocean. The source of water here is rainfall. The water percolates into the crust where it is heated by subsurface magma and then rises back to the surface. Because of the composition of the local rocks and the nature of the area's 'plumbing', many of the hot water vents are in the form of geysers. Others occur as hot springs or warm springs, and the deposition of minerals at the surface by the hot water leaves behind fascinating structures (Figure 3.10).

[Figure 3.10]
Terraces of the calcium-rich mineral travertine deposited from hot springs that are releasing water at the surface, at Yellowstone National Park. The individual terraces are about 20 cm high, and are filled with hot water that is flowing across them.

Plate tectonics back through time

Not only can we see that the plates at the surface of the Earth are moving today, but we can trace their movements back in time. This is done using measurements of the history of the magnetic field, which tells us about the changing locations of different regions, and geological evidence that tells us the history of individual regions and the ages of the rocks. The movement of the continents and of the plates on which they are sitting can be traced back in time for about a half billion years. Beyond that, there is not enough information to decipher fully what was happening. About 200 million years ago, all of the continental material was located in one giant supercontinent, termed Pangaea. Pangaea was created by the collision of two earlier large continents, termed Laurasia and Gondwanaland.

Laurasia included what is now North America, Europe, and Asia, and
Gondwanaland contained South America, Africa, India, Australia, and
Antarctica. Rifting split Pangaea into two masses, and these two masses
subsequently split apart and the continents moved to their present locations.

Of course, the continents are continuing to move, as evidenced by the
ongoing creation and destruction of crust and by earthquakes around the
world. The continents are moving apart at an average rate of about 10
cm/year; at this speed, it would take only 200 million years for the continents to move from one side of the Earth to the other. Clearly, in another 50
million years, the global distribution of continents will be extremely different from what it is today.

As the continents move in response to the mantle movement, new sea
floor is continually being created and old sea floor is being destroyed. As a
result, the oldest oceanic crust that can be found is only a couple of
hundred million years old. All of the older crust has already been consumed. The Earth truly is a dynamic planet.

What was the nature of plate tectonics during the first half of the Earth's
history? We expect that mantle convection and plate tectonics has been
occurring throughout the entire history of the Earth. The oldest known
direct indication that plate tectonics occurred, however, comes from
seismic observations in the Superior province of Canada. These suggest
the presence of an ancient subduction zone there that would have been
operating about 2.7 billion years ago. At that time, the abundance of radioactive elements was greater (they had not yet decayed to the present level),
so the Earth's interior would have been hotter and mantle convection
should have been more vigorous.

In addition, geochemical analysis of rocks from the mantle suggests
that some significant event of global proportion happened about 2 b.y.a.
Although the data do not indicate what the nature of that event was, it
might have been the global onset of plate tectonics and the accompanying
beginning of mixing of rock between the mantle and crust.

It is possible that farther back than 2–3 b.y.a. the cool, brittle region
near the surface was too thin to be effectively pushed beneath the surface
and be subducted. However, this does not mean that mantle convection
was not occurring at that time. The fact that there would have been more
heating of the Earth's interior at that time suggests that the mantle must
have been convecting even more rapidly than it is today in order to eliminate the heat. If plate tectonics was not occurring then, this just means
that the surface expression of the mantle convection must have been
different than at the present. All of the large planets in the inner solar

system (Earth, Venus, and Mars) are thought to have convecting mantles, based on the expected rates of heat production in their interiors, but only the Earth expresses this at the surface by the occurrence of plate tectonics. This will be discussed later, when we talk about the other planets in our solar system.

Early history of the continents and oceans

Finally, let us discuss the origin of the continents and of the oceans. The oldest continental rocks are about 4.0 b.y. old, and the oceans may be this old as well. Crustal material has probably been forming throughout almost the entire history of the Earth.

Although both the continental and the sea-floor materials are called crust, they have very different compositions and were formed by very different processes. The Earth's mantle is made up primarily of olivine, a green mineral that contains atoms of silicon, oxygen, iron, and magnesium; there are other minerals mixed in with the olivine, however. Volcanism occurs when some of this rock is melted and the melt is pushed to the surface. The molten rock that rises up to the surface will cool off, and crystals of the minerals olivine, pyroxene, and feldspar will form. This change from a predominantly olivine composition arises because the first minerals to melt on heating the mantle are those that melt at the lowest temperatures, so that a melt created by only partly melting mantle rock will not contain all of the original minerals. As the melt cools off and solidifies, therefore, it forms slightly different minerals than were initially present. The particular mixture of minerals that results from partly melting mantle rock is called basalt, and makes up the crust that forms at mid-ocean spreading centers.

When basalt on the ocean floor is pushed into the upper mantle at subduction zones, it, too, will be heated and will partially melt. This new magma rises to the surface, produces volcanoes, and cools off. However, partially melting basalt again will melt only some of the minerals. When this melted rock comes to the surface and cools off, it will produce slightly different minerals. Recycling the basalt in this manner will produce rock that contains quartz and feldspar but little or no olivine or pyroxene; this combination of minerals is known as granite. The granite is not as dense as the basalt, so it will tend to stay at the surface rather than subduct. It eventually accumulates to form the continents. And, being both less dense and thicker, the continental rock will float higher above the mantle, so the continents stand higher than the ocean floor. A good example of the formation of continental crust would be the Andes Mountains along the western coast of South America. They formed as a result of the subduction of the Pacific plate beneath the continent. Another example is the Rocky Mountains that resulted from the subduction of the Pacific plate beneath North America.

Because the continental crust is created by the processing of subducted oceanic crust, the creation of new continental crust has occurred through-out time and much less continental crust was present during the first half of the Earth's history. In fact, a surge of creation of new continental material a couple of billion years ago may have coincided with the onset of plate tectonics. Continental crust is continuing to form, primarily at sub-duction zones where basalt is processed to make granite.

The timing of the formation of the oceans and atmosphere, clearly of relevance to understanding the origin and evolution of life, is related to the other geological processes that were occurring on the Earth. As the Earth accreted 4.5 b.y.a., the individual impacts that brought material onto the surface of the Earth probably released gases into the atmosphere. In addi-tion, the Earth would have formed a core very early in its history, with the heat from the impacts melting the outer parts of the planet and allowing iron to sink to the middle. This core-formation process would have released enough heat to melt the entire Earth, and some of the gases that were held within the interior would have been released as well.

There is evidence from different isotopes of the atmospheric gases that at least some of the gases were released from the interior very early. The isotopes are different versions of the same atom, having different numbers of neutrons in the nucleus of the atom. For example, argon-40 and argon-36 both are atoms of the noble gas argon. Each has 18 protons in the nucleus, so they both behave the same way chemically. Argon-36 has 18 neutrons in the nucleus as well, and argon-40 has 22 neutrons (with the '-36' and '-40' designating the sum of the number of neutrons and protons). As a noble gas, argon does not react chemically with other species; this means that once it is put into the atmosphere it will stay there. Argon-36 is the naturally occurring stable isotope of argon, while argon-40 is produced by the radioactive decay of potassium-40. The decay of potassium-40 occurs with a half-life of 1.25 billion years, meaning that half of the potas-sium-40 that is present will decay to argon-40 in 1.25 b.y. As a result, the abundance of argon-40 in the atmosphere tells us about the timing of out-gassing – if the release of gas from the Earth's interior happened very quickly after the Earth formed, then there would not have been enough time to produce very much argon-40 and its atmospheric abundance would be quite small. If outgassing occurred very late in history, however, there would have been sufficient time to produce a lot of argon-40. The relatively small abundance of argon-40 actually present in the atmosphere suggests that much of the outgassing of the Earth occurred very early in its history.

The isotopes of xenon, another noble gas that will tend to remain in the atmosphere, tell us the same thing but for a different reason. The different isotopes of xenon, none of them radiogenic, show a degree of loss that depends on the mass of each isotope. This suggests that there has been a loss of xenon from the atmosphere, with the lighter isotopes having been removed preferentially. The only mechanism that can remove xenon from the atmosphere involves the loss to space of vast quantities of hydrogen. As the hydrogen escaped, it would drag some of the xenon along with it; the lighter isotopes would have been easier to drag along, and would have been preferentially removed by this process. The rapid loss of substantial amounts of hydrogen can have occurred only during the Earth's earliest history, as a part of the accretion of the planet, when substantial quantities of hydrogen were released from the accreting material. This means that some xenon must have been in the atmosphere during the earliest history of the Earth, and again suggests that at least some of the planet's gases must have been released into the atmosphere very early.

The geological evidence also suggests that the oceans already were present at the time that the earliest crustal rocks were created, about 3.8–4.0 b.y.a. The oldest crustal rocks that can be found are metamorphic rocks, which were created by the application of heat and pressure to previously formed sedimentary rocks. Deposits of sediment that can be transformed into rocks form at the bottom of bodies of water, so the fact that sedimentary rocks had existed by that time shows that at least some fraction of the current oceans also must have existed. Also, deposits of carbonate rocks exist that date back to this same time period. Carbonates are made of minerals such as limestone, which contain carbon dioxide gas combined with various elements (such as calcium). Today, limestone forms from the shells of ocean creatures, but it can also form directly by atmospheric carbon dioxide dissolving into the oceans and then precipitating out as carbonate mineral. Because carbonates have to be formed under water, their presence 3.8 b.y.a. again tells us that the oceans existed at that time.

The composition of the carbonates suggests that there must have been abundant carbon dioxide gas present in the atmosphere at the same time. The CO_2 gas would have trapped some of the Sun's heat by the greenhouse effect and warmed the surface. Surprisingly, there may have been enough

CO_2 in the atmosphere that the temperature at the surface of the Earth could have been as high as 85 °C.

The atmosphere has been evolving continuously over the last 4 billion years. Today, most of the Earth's CO_2 is present as carbonate (limestone) deposits. But, this CO_2 can be recycled back into the atmosphere – sediments, containing CO_2 and water, are subducted beneath the surface at subduction zones, the water and CO_2 are released from the rock as it is heated up, and they are put back into the atmosphere through volcanoes. There is also some continued release of fresh, non-recycled gas into the atmosphere from the deep interior. This outgassing occurs in association with volcanism at mid-ocean ridges and at hot spots. And, as we will see later, living organisms have affected the atmosphere substantially, being responsible for the production of the oxygen that is present today.

Concluding comments

Clearly, the very early Earth was not the Earth that we know today – one where temperatures are relatively clement, where continental material is widespread, and where geological processes seem to occur at a relatively bucolic pace. Rather, the early Earth must have been a much more dynamic place. Ongoing impacts perturbed the climate substantially, providing new gases to the atmosphere. The oceans were present by about 4.0 b.y.a., but it is likely that little or no continental material was present. Volcanism would have been much more active than it is at present, although much or all of it might have been invisible beneath the ocean's surface. Plate tectonics might not have been occurring, despite the more vigorous convection of the mantle; the crust might have been thin, and hot-spot volcanism might have played a more important role in getting rid of heat. Atmospheric temperatures might have been substantially higher than they are today, possibly as high as 85 °C.

Most importantly, the Earth that existed 4 billion years ago probably contained no life, at least not that was able to survive up to the present. It may be hard to imagine the continents and the ocean floor without life, but remember that at that time there were no continents and the ocean floor may have been much more volcanically active! Only over perhaps several hundred million years did life originate and spread throughout the world.

54 Most of the important clues we have regarding the history of life on Earth come from the geological record as read from rocks of different ages. These include the antiquity of life, dating back to at least 3.5 billion years before the present; the small size and simplicity of the earliest cells and their evolution into more complicated cells; the nature of the environment in which early life flourished and the changes that have occurred through time; the dramatic increase in the size and complexity of many species about 600 million years ago; and the subsequent history and evolution of life up to the present. In order to understand the constraints that these clues place on the origin of life, we need to understand the nature of the record of the processes that have occurred and the nature of the record of life that exists within the rocks.

When we look at the rocks that we find around us, we typically can see layer upon layer of different rock types. These can be seen easily in canyon walls, for example, where a stream or river has carved down through the rocks to expose the different layers. These layers tell us about the processes that have shaped the surface. The rocks themselves can tell us about the climate and environment that existed when they were formed, or about the geological processes that were taking place then. The fossils that are found within the layers tell us about the life that existed at the time that the rocks were formed.

Typically, sediments are deposited on top of any preexisting rocks and can subsequently be turned into rocks themselves. This means that the new layers are deposited horizontally. By digging deeper down through the strata, we are accessing older rocks, and we can learn about the processes that occurred in earlier times. For example, the walls of the Grand Canyon in the southwestern U.S. show layers of rock that were deposited over a 2 b.y. period; the rocks on the top are 2 b.y. younger than the rocks on the bottom. (The mechanisms for determining the ages of rocks will be described below.) At one place or another on the Earth, we can find rocks that date back to just about every epoch in the Earth's history since about 4.0 b.y.a. This allows us to reconstruct the basic history of the Earth's surface for that entire time period. Even though the Earth formed about 4.5 b.y.a., no rocks have been found that were formed before 4.0 b.y.a. and the earliest half-billion years remains a mystery. Given the discussion in Chapter 2 regarding the rate of large impacts onto the early Earth and the dynamic nature of the planet, however, it is not surprising that no rocks remain intact from this period.

The rocks that tell us about the history of life are, for the most part, sedimentary rocks. They formed by the compaction and solidification of sediments such as sand or silt that typically were deposited under water.

For example, sediments are carried into lakes or the oceans by fast-moving streams. There, the water slows down, and the sediment will settle out of the water and accumulate onto the lake or ocean floor. Over time, layer upon layer will be deposited. Once solidified by compaction, these layers can be exposed by tectonic uplift or by further erosion by streams. The top layers clearly were deposited more recently than the bottom layers; this allows us to infer the relative ages of the different deposits (in other words, which layers are the youngest and which are the oldest).

As the layers of sediment form, they can incorporate within them any organisms that are present at the time. These creatures might have died and sunk to the bottom or, in the case of quickly deposited sediments, not been able to move fast enough to get out of the way. The flesh of animals usually decays away quickly, but the bones or shells remain behind; these can become fossilized and remain trapped within the rock until exposed by erosion. Not surprisingly, perhaps, we find that different layers contain fossils from different species. In fact, the types of species that are found within a given layer are very consistent from one region to another. The presence of different fossils occurring in different layers shows a change over time in the species that existed on the Earth, and allows us to reconstruct the history of life on Earth. This history is remarkably consistent from one location to another, spanning the entire world (Figure 4.1).

This is an important point that cannot be overemphasized: the geologic record provides clear and convincing evidence that the species living on the Earth have changed through time. These changes are recorded as fossils that have been incorporated into rocks of different ages, and the ages of the rocks tell us when each different species existed. By no means have we catalogued and described every species that has existed through time, but we can look at related species, for example, and see how they have evolved.

The changes in the species that are found within layers are often so notable that we use these changes to define different periods in the Earth's history. For example, dinosaur bones are found only in layers of rock that were deposited between 240 and 65 m.y.a.; this time period constitutes what is called the Mesozoic era in Earth's history. The extinction of the dinosaurs 65 m.y.a. (along with a large number of all species) is seen in the geologic record as the absence of dinosaur fossils in all younger rocks. The change in the types of fossils found within the rocks 65 m.y.a. is so marked that we use it to define the end of the Mesozoic era and of a subdivision of the Mesozoic called the Cretaceous period (see Chapter 2).

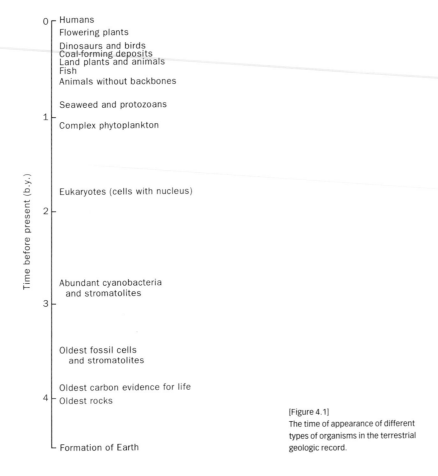

[Figure 4.1]
The time of appearance of different types of organisms in the terrestrial geologic record.

No fossils of species with hard parts such as bones or shells are found in rocks older than about 570 m.y. The era from 570 m.y.a. up to the present is called the Phanerozoic era (from Greek words meaning, literally, the 'era of visible life'); the oldest part of this era is called the Cambrian period. Rocks older than 570 m.y. contain very few fossils that can be seen without a microscope. In fact, this pre-Cambrian time period once was thought to have contained no life whatsoever. As we will see, however, there are fossils of living species in the older rocks, dating back 3.5 b.y.

The techniques of field geology provide information on the relative ages of the different rocks. This approach involves looking at the relative stratigraphy (the sequence of what rocks lie atop what rocks) and the succession of fossils throughout the layers, and allows a determination of what the order was in which the rocks were created. However, this does not

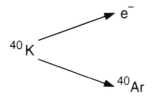

[Figure 4.2]
Schematic diagram showing the radioactive decay of potassium-40 to argon-40 by the emission of an electron.

tell us the actual date of the rocks' formation. Despite this, geologists during the past century had a sense of the rates at which different geological processes acted. This sense was based on their observations of how these same processes act today, and of how long those processes would have taken to form the different rock sequences and geological features that are seen. Even though all of the estimates suggested a planet that had existed for hundreds of millions of years, the estimates also varied by many hundreds of millions of years. In order to determine the absolute ages with some accuracy, however, a different technique must be used. Today, absolute age dating is done by the use of radioactive elements.

Radioactive atoms are capable of changing from one element to another by various nuclear decay processes. These processes were discussed briefly in Chapter 3, because they give off minute amounts of heat that drive the heat engine of the Earth's interior. Here, we are interested in the fact that the rate at which this decay occurs is constant and can be measured in a laboratory. Then, we can measure the number of atoms of a radioactive element that are in a sample of a rock, along with the number of atoms into which this element will decay, and use this to estimate the age of the rock. For example, an atom of radioactive potassium will decay by emitting an electron to become an atom of the non-radioactive gas argon (Figure 4.2). In this case, the more argon that is present within a rock and the less radioactive potassium, the longer the time period that must have passed since the formation of the rock.

This technique works because, when a mineral grain forms, it will incorporate small amounts of the potassium as a contaminant; however, the isotope of argon that is produced by the decay of potassium (argon-40) does not otherwise occur in nature, and argon molecules previously formed by decay will not be included within the mineral. Once formed, the mineral will keep trapped within it any argon created by the decay of potassium. Therefore, the relative amounts of potassium and argon tell us the number of years that have passed since the mineral grain formed. Usually, this is the number of years since the mineral solidified from molten rock and first became able to hold argon within it. This technique is an excellent method for determining the age of igneous rocks, since they formed by crystallization from molten rock. It is not as useful for determining the ages of sedimentary rocks, however, since sediments were created by incorporating individual mineral grains that themselves were formed at an earlier time.

A variety of different radioactive 'clocks' can be used to measure the ages of rocks. In addition to the potassium–argon technique described above, rubidium decays into strontium and can be used to determine age, and uranium decays into lead and can also be used. Each of these combinations of 'parent–daughter' isotopes works better in some situations than others, as long as one keeps in mind the limitations of each technique.

By using these absolute age-dating techniques, we can determine the dates of various geologic events. We know that the oldest rocks that have been found are 4.0 b.y. old, and that the youngest rocks found are 0 years old (of course, we know this last one because rocks can be seen forming by the solidification of molten rock in lava flows at volcanoes, for instance in Hawaii). By combining the relative stratigraphy determined by field investigation of the rock strata with laboratory determinations of the ages of those rocks which can be dated, we can construct a history of the geologic processes and of the evolution of life throughout Earth's history. (We also know that the age of the Earth itself is 4.5 b.y. This comes from treating the entire Earth as a unit and looking at when it became segregated from the meteorites.)

The history of life and climate on Earth

At this point, the standard discussion of the history of the Earth would use the evidence from the geologic record to trace the history of life from its earliest appearance up to the present. Such discussions comprise what is called historical geology. Our main interest here, however, is to understand the evidence for the earliest occurrence of life on Earth. Therefore, we will start with the most recent occurrences, as they will be more familiar to us and easier to understand, and work backward in time. By comparison, we started with the formation of the Earth and worked forward in time in Chapter 2. These two approaches both lead us to the time period between about 4.0 and 3.5 b.y.a., when life must have originated. Although we will see that there is no direct evidence that tells us about the origin of life, each of these approaches provides significant constraints on both the processes and the timing of the origin.

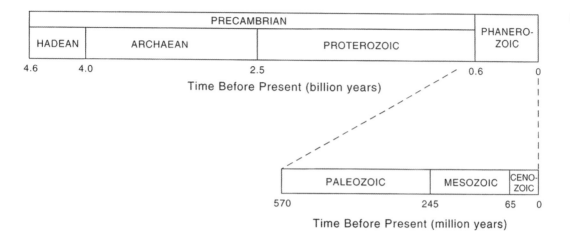

The history of the Earth is usually divided up into three eons (Figure 4.3) – the Archaean, from the time of the oldest rocks, 4.0 b.y.a., up to 2.5 b.y.a.; the Proterozoic, from 2.5 b.y.a. up until 570 m.y.a.; and the Phanerozoic, from 570 m.y.a. up to the present. The time from the formation of the Earth up to the oldest known rock often is referred to as the Hadean eon due to the 'hellish' nature of the environment.

The event that marks the beginning of the Phanerozoic eon 570 m.y.a. is the appearance within the geologic record of substantial diverse fossils. Prior to this eon, geological evidence for the nature of living species is relatively rare, and was recognized only within the last few decades. At the start of the Phanerozoic, with a period labeled the 'Cambrian', numerous different forms of life are seen in the fossil record (Figure 4.3). These fossils, which appeared to arise from no previous life forms, show the basic diversity and styles of all subsequent life. The diversity of animal fossils represents all of the basic types that we see today, including, for example, insects, crustaceans, and chordates (animals with a spinal cord, of which humans are one species). Not all species were present at that time, and not even all families of species. For example, fish do not appear within the geological record until several tens of millions of years later, and the first land plants do not appear until 100 m.y. later.

The sudden appearance of the many different types of species at the beginning of the Cambrian period, and the very rapid spread of life into different ecological niches and different species has been called an 'explosion' of life that began 570 m.y.a. New species formed rapidly, and evolution of these species was widespread.

In addition, of course, existing species also became extinct over time.

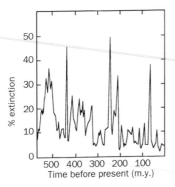

[Figure 4.4]
The percentage of all genera to go extinct during each time interval during the Phanerozoic epoch (since about 600 m.y. ago). There are five major extinction events; the two most recent events have impacts associated with them, and the next previous two may have impacts associated with them. The geologic record is incomplete, and there may have been additional impacts that were large enough to cause an extinction event. (After Rampino and Haggerty, 1994.)

Not all species became extinct, of course, but, over time, most of the species that have existed on Earth have died out.

During the last several hundred million years, in fact, there have been numerous times when a large fraction of species suddenly became extinct (Figure 4.4). These so-called 'mass extinction' events were mentioned in Chapter 2. The cause of these extinctions may have been sudden changes in the climate. The underlying cause for the climate change may have been increased volcanism or, in some instances, the impact of a large asteroid or comet onto the Earth's surface. In the largest extinction event, marking the end of the Permian period about 250 m.y.a., close to 95% of all species disappeared. From the 'evolutionary' perspective, the significance of these extinctions was that many ecological niches suddenly became available. The species that had previously occupied them were no longer alive, and new species able to take advantage of these niches could evolve. If there were a place in the environment where a living creature could benefit, in terms of its place in the food chain, for example, a species could rapidly evolve to do so.

A specific example of species evolving to take advantage of these new ecological niches occurred at the end of the Cretaceous period, when the dinosaurs became extinct. Prior to the demise of the dinosaurs, mammals were relatively small species that by no means dominated the environment. Subsequently, however, these small mammals were able to evolve into larger species and to take advantage of the absence of dinosaurs. They have since become one of the significant types of beings on the surface of the Earth. In many respects, it was this singular and unpredictable event – the extinction of the dinosaurs – that led some of the species of mammals to begin moving down a path that eventually would allow some of them to evolve into humans some 65 m.y. later.

Over the last several decades, there has been the recognition that life did not arise at or near the beginning of the Cambrian period. In particular, a wide variety of macroscopic species existed during the several tens of millions of years before the Cambrian began. Of most interest are those species that were initially found fossilized in a particular layer of rock called shale found in British Columbia, Canada. This 'Burgess shale', as it is called, contains rare fossils of soft-bodied creatures that existed before the hard-bodied species of the Cambrian. What may be the most significant aspect of the creatures in the Burgess shale is that they show a variety of body forms that are very different from those seen in the Cambrian and in subsequent life. That is, the body types were much more diverse. Most died out, however, leaving relatively few different body types to evolve into all subsequent macroscopic life. There is no apparent or obvious evolutionary advantage noticeable in the species that survived, and it is thought that the selection of surviving species was somewhat random and arbitrary. As Stephen Jay Gould, a paleontologist at Harvard University, points out, if one were to rewind the tape of life and play it over again, and let events occur randomly as they must have the first time around, there is no reason to believe that life would take the same evolutionary path. Life would evolve, but the path of evolution would be a different one. We will return to this topic in the discussion in Chapter 17 of whether extraterrestrial intelligence is likely to exist.

Life did not originate near the end of the pre-Cambrian epoch. Nor did multiple-celled life originate at that time. The oldest evidence for multicellular life consists of fossils that contain carbon within them and appear to be leaf-like in structure. These date back to 1.7 b.y.a.

The period from 2.5 b.y.a. to the end of the pre-Cambrian about 570 m.y.a. is called the Proterozoic era (from the Greek, meaning 'era of earlier life') (Figure 4.3). The boundary between the Proterozoic and the earlier Archaean is somewhat arbitrary, but the division into two different eras reflects the changing nature of the environment. During the Archaean, the atmospheric CO_2 abundance was high and the O_2 abundance was low; during the Proterozoic, CO_2 was substantially reduced and O_2 was present in significant amounts. In addition, around the time of the boundary between the two eras, the production of continental crust underwent a dramatic increase; much of the world's present continental material dates back to this time.

The change in the oxygen level is one of the significant aspects of the middle history of the Earth because of what it tells us about the evolution of oxygen-producing, photosynthetic life. The evidence for a shift in oxygen level can be seen in the relative abundances of oxygen-rich and oxygen-poor (oxidized and non-oxidized) rocks in the geologic record (Figure 4.5).

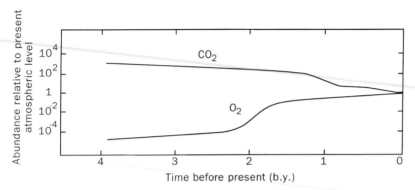

[Figure 4.5]
The history of the atmospheric abundances of CO_2 and O_2 in the Earth's atmosphere. Values are shown relative to their present atmospheric level. At earlier times, there are large uncertainties in the values, although the trends probably are approximately correct. (After Schopf, 1992.)

For example, consider what are called 'banded-iron formations'. These rock units contain layers that consist of at least 15% oxidized iron (iron combined with oxygen); the banded nature comes from the presence of layers of iron-rich rock intermixed with layers of iron-poor rock. Typical layers may be only centimeters thick, but the thickness of whole sequences can be up to hundreds of meters thick. Most of the banded iron-formations were deposited between 2.5 and 2.0 b.y.a. Where it is feasible, these formations are often mined for their iron. The major process by which banded-iron formations occur is thought to be oxygen combining with iron dissolved in the oceans. Iron that has not been fully oxidized can dissolve readily in seawater; the source of the iron may be either erosion from the continents or volcanic eruptions at sea-floor spreading centers. When the dissolved iron combines with oxygen, it becomes insoluble in water; at that point, it precipitates into a solid and settles onto the ocean floor. During the Earth's early history, there must have been abundant dissolved iron in the oceans. Whenever oxygen became available in the atmosphere, it too would dissolve into the oceans where it would combine with the iron and form sediments. The abundance of iron formations around 2 b.y.a., then, logically represents a time at which the abundance of oxygen in the atmosphere increased. Banded-iron formations subsequent to this time are rare, presumably due to a lack of abundant unoxidized iron rather than to a lack of oxygen.

Additional evidence for the change in the abundance of atmospheric oxygen comes from deposits of the minerals pyrite (FeS_2) and uraninite (UO_2). Both of these minerals would combine very easily with oxygen if it were present, and produce different minerals. Deposits of these minerals that were formed between 2 and 3 b.y.a. contain the unoxidized minerals. The individual grains are rounded, indicating that they had been transported in an environment in which erosion could occur, such as in a stream bed. If oxygen had been present, it would have dissolved in the water and come into contact with the mineral grains and combined chemically with them, thereby oxidizing them. Again, there is an absence of these deposits from more recent times, suggesting that the oxygen levels have increased.

The rise in oxygen level generally is thought to have come from the production of oxygen by photosynthesis in bacteria and plants. Although oxygen can be produced in small quantities by photochemical processes whenever CO_2 is present in the atmosphere (as in the present-day Venus and Mars atmospheres, for example), the amount of oxygen produced by non-biological mechanisms is thought to have been relatively small. The build-up of oxygen around 2 b.y.a. probably does not indicate the time for the onset of photosynthesis but, rather, that all of the available chemical sinks for oxygen had become exhausted. These sinks included the oxidation of iron and other minerals in the oceans, for example.

At the same time that the atmospheric oxygen levels were increasing in the Earth's atmosphere, the carbon dioxide levels were decreasing (Figure 4.5). CO_2 is continually being supplied to the atmosphere from the Earth's interior through volcanic eruptions. In fact, the amount of CO_2 in the present-day atmosphere could be supplied in only 400 000 years. Clearly, CO_2 must be undergoing a steady removal from the atmosphere in order to maintain a level near its present value, and the amount in the atmosphere must represent a balance between supply to and loss from the atmosphere. Although biological activity is a significant player in this CO_2 cycle (with photosynthesis utilizing CO_2 and subsequent consumption of plants by animals and the oxidation of their cells producing CO_2), the main player in this balance appears to be chemical reactions in the oceans. Atmospheric CO_2 dissolves into the oceans, where it reacts with other minerals to form deposits of carbonate minerals. The carbonate deposits rest on the ocean floor and, ultimately, are carried back into the mantle at subduction zones. As the subducted rocks are pushed deeper into the mantle, some of them will melt and produce the volcanism that is associated with subduction zones. This volcanism resupplies the CO_2 back to the atmosphere, completing the cycle (Figure 4.6).

It is the rate of formation of the carbonates that determines how much CO_2 will be present in the atmosphere. A faster carbonate formation rate, for example, will pull more gas out of the atmosphere and deposit it as

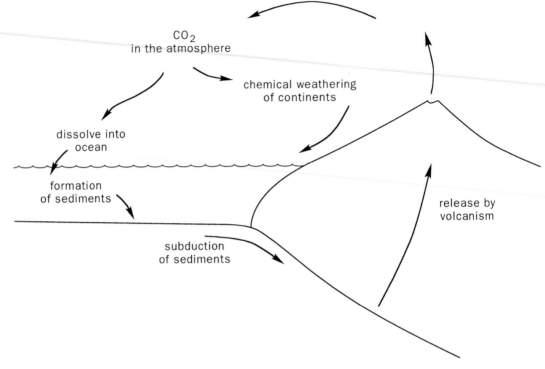

CO$_2$
in the atmosphere

chemical weathering
of continents

dissolve into
ocean

formation
of sediments

release by
volcanism

subduction
of sediments

[Figure 4.6]
Schematic diagram of the terrestrial CO$_2$ cycle. CO$_2$ is released into the atmosphere through volcanism. From there, it dissolves in the oceans and precipitates to form sediments; these sediments are subducted back into the mantle, and the CO$_2$ is released again through volcanism. The various feedbacks between atmospheric temperature and CO$_2$ abundance may serve to control the amount of CO$_2$ in the atmosphere and to keep the surface temperature clement.

minerals on the ocean floor; this will reduce the amount of CO$_2$ that is present in the atmosphere. Conversely, a slower rate of carbonate formation will allow more CO$_2$ from volcanism to build up in the atmosphere. Since the rate at which carbonates form is very sensitive to the ocean temperature, a simple feedback mechanism can be set up: if more CO$_2$ is supplied to the atmosphere by volcanism than is removed by carbonate formation, then CO$_2$ will build up in the atmosphere. The increase in the amount of CO$_2$ in the atmosphere will cause the temperature to increase due to greenhouse warming, and the rate of carbonate formation will increase as a result. This has the effect of removing the excess CO$_2$ in the atmosphere and bringing the temperatures back to normal. On the other hand, if CO$_2$ is removed from the atmosphere faster than it can be supplied, then the CO$_2$ abundance in the atmosphere will decline. As a result of the decrease in CO$_2$ and the resulting decline in the efficiency of the atmospheric greenhouse, atmospheric temperatures will drop and carbonate production rates will drop. This will then allow CO$_2$ to build up again in the atmosphere. The net result of these competing processes will be a balance of temperature, atmospheric CO$_2$ abundance, and carbonate formation rate. If any one of these three properties is pushed away from its equilibrium value, the other two will respond in a way that tries to bring the system back to the equilibrium.

The present-day atmospheric abundance of CO_2 is about 350 parts per million parts of atmospheric gas (that is, about 3.5/100 of one per cent). This is a very small fraction of the total available CO_2 on the Earth; most of it is present as carbonates. If all of the carbonates were converted back to CO_2 and placed into the atmosphere, it would be the equivalent of about 60 bars of atmospheric pressure (one bar is the atmospheric pressure produced by the entire atmosphere, so this is close to 170 000 times the amount of CO_2 present in the atmosphere today). The recycling time of atmospheric CO_2 is 400 000 years, so the atmosphere and climate would respond to any changes on this timescale – it would take this long to establish a new equilibrium, for instance, if the rate of volcanism changed.

There are few indicators in the geological record that can tell us what the CO_2 level was at any given time. The oldest rocks in the record are metamorphic, but were formed from sedimentary rocks about 4 b.y.a., requiring that sediments could be deposited in water; the presence of liquid water requires that the temperature was above the freezing point of water, 0 °C. Since the intensity of the Sun's output was lower than its present value by nearly 30% in its earliest history, the temperatures would have been dramatically lower, less than freezing, unless a greenhouse gas were present; CO_2 is the most likely gas (and possibly the only one available, as other gases such as methane or ammonia may not have been stable). In fact, the CO_2 abundance would have to have been several hundred times greater than its present level in order to have raised the temperature to above the freezing point; the CO_2 pressure may have been as high as 10 bars. Interestingly, if the CO_2 pressure was near this high value, surface temperatures would have been around 85 °C – the temperature of hot coffee.

By 2.5 b.y.a., the temperatures must have been low enough for glaciation to occur at least occasionally, in order to be consistent with the geological evidence for glaciers, but not so low that the entire planet would freeze. Temperatures in the range of 5 to 20 °C are expected to have been typical, consistent with a CO_2 abundance between about 50 and 500 times the present level (Figure 4.5). As the solar luminosity continued to increase to its present value, CO_2 levels would have continued to decline. The increasing intensity of the Sun through time, combined with the ability of the CO_2 in the atmosphere and in carbonates to respond to the temperature, may have resulted in the temperature being maintained at a nearly constant level from the middle history of the Earth all the way up to the present.

66   The earliest evidence for life on Earth

Now that we have a sense of the history of the Earth's climate and life through time, we can discuss the earliest evidence for life, during the Archaean era. As we will see, convincing evidence exists for the existence of life 3.5 b.y.a.; less convincing, but still plausible and intriguing, evidence suggests that life might have existed as long ago as 3.85 b.y.a.

The geological evidence tells us what the environment was like at the time of the earliest known life: the oldest rocks, dated to be 4.0 b.y. old, include some that were deposited as sediments; this tells us that the oceans existed at that time and that temperatures were above the freezing point of water. Although oxygen was not present in significant quantities, the occurrence of carbonate mineral deposits tells us that carbon dioxide was present. Very little continental crust had been formed, however; there might not have been any land sticking up above the ocean surface, except perhaps on the rims of giant impact craters.

There are three lines of evidence that tell us about the earliest known life – the occurrence of structures that appear to be stromatolites, consisting of colonies of bacteria that are similar to those that are seen to occur today and are clearly of biological origin; the existence of microscopic fossils within sedimentary rocks that are identical in form to modern-day single-celled life; and carbon isotopic evidence that may suggest the existence of life as long ago as 3.85 b.y. It is important to keep in mind that the evidence we will discuss is the oldest known evidence for life. It is not necessarily evidence that is relevant to the very first life that existed on the planet, and it does not tell us about the origin of life itself. More than likely, life that formed these oldest fossils had evolved substantially from what must have been the initial life, and it had already become quite sophisticated by the time these clues were locked into the geologic record. Also, it is important to realize that we are looking for clues in a geologic record that has many holes; as such, we cannot expect the evidence for life to be entirely unambiguous.

The first clue as to the earliest life comes from ancient deposits that are thought to be analogous to present-day stromatolites. Stromatolites are macroscopic structures built up by the growth of bacteria. They incorporate sediments that are deposited on top of them in shallow water, and have a layered structure as a result. Living stromatolites are relatively rare today, primarily because the colonies of bacteria serve as food for certain animals; they can be seen, for example, at Shark's Bay in Western Australia.

The stromatolite structure consists of layers of different types of bacteria. The bacteria at the top can exist in (and make use of) oxygen in the atmosphere, and they can use sunlight to drive photosynthesis. Beneath these are bacteria that can use oxygen if it is available but can also produce energy by fermentation when there is no oxygen. On the bottom, at a depth to which no oxygen can diffuse without first being used up, are the bacteria that cannot survive in the presence of oxygen. The first two layers may be only a millimeter or two thick, while the lowermost layer can be substantially thicker. As sediments are deposited on top of the stromatolites, the bacteria migrate up through the sediments in order to maintain their proper depth. In this manner, the sediments are left behind in discrete layers, and layer upon layer can be built up. The structures that are left behind typically are tens of centimeters in size.

Fossil structures that appear identical in size, shape, and interior structure to the stromatolites are found throughout Earth's history (Figure 4.7). There was some disagreement initially as to whether they were formed by biological or nonbiological processes until the living stromatolites were discovered. It is possible, however, that some of the structures might have formed by nonbiological processes; these might involve deposition of layered sediments and subsequent slumping of the sides, for example. As a result, stromatolite structures by themselves are not unique evidence for the existence of life, but they are suggestive. The oldest stromatolite structures are found in Swaziland, South Africa, and in the Pilbara rock formations in Western Australia. Both of these deposits are between 3.3 and 3.5 b.y. old.

The second line of evidence regarding the earliest life consists of individual fossilized cells that date back to 3.5 b.y.a. These cells have been petrified within sedimentary rocks that were deposited at that time; they can be seen in abundance in 'thin section' slices of rock, examined under a microscope. The individual cell structures have been determined to contain organic carbon, and have an appearance very similar to various types of bacteria that are alive today (Figure 4.8). Some of the fossil cells show features such as cell walls and what appear to be the fossilized remnants of DNA molecules. Typical sizes for these cells are less than 50 microns across (there are 1000 microns in a millimeter), similar to the size of those living cells that have identical appearance. Some of the fossil cells are spherical in shape and others have more elaborate shapes, again similar to living cells. The fossils that are most convincingly of biological origin are those that are filamentous in shape and occur in association with fossil stromatolites.

Some of these fossil cells are most similar to modern cyanobacteria. These bacteria obtain their blue-green color from the presence of chemicals such as chlorophyll that drive photosynthesis – the production of

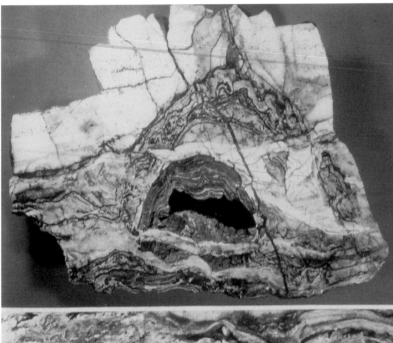

[Figure 4.7]
Domical (top) and flat-laminated (bottom) stromatolite-like structures, found in rocks that are 3.5 b.y. old in the Warrawoona Group in Western Australia. The domical structure is about 30 cm across, and the flat structure is about 10 cm across. (Photo courtesy of J.W. Schopf.)

usable energy directly from sunlight. If the fossil cells really were photosynthetic, it can be argued that relatively advanced life (from a biochemical perspective) existed as early as 3.5 b.y.a. It is possible, of course, that bacteria existed prior to 3.5 b.y.a., but the geologic record from these earlier years is extremely sparse.

The final argument is based on the ratio of the different isotopes of carbon in the sedimentary rocks. Carbon has two stable isotopes – carbon-12, the most common, has 6 protons and 6 neutrons in its nucleus; and carbon-13 has 6 protons and 7 neutrons, so is a little bit heavier. Although both isotopes will participate in the same chemical reactions, the lighter

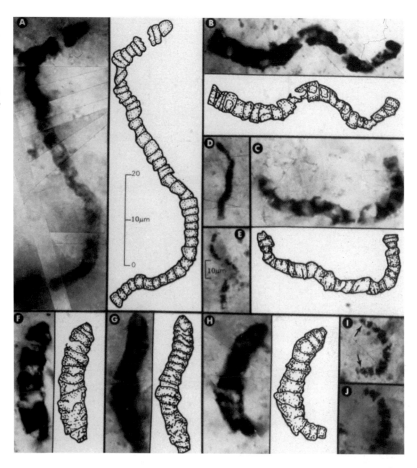

[Figure 4.8]
Fossil bacteria found in 3.5 b.y. old chert in Western Australia, shown along with interpretative drawings. Images D, E, I, and J are at the scale shown in E; all others are at the scale shown in A. (Figure courtesy of J.W. Schopf.)

mass of carbon-12 causes it to be slightly preferred in the chemical reactions that occur within living organisms. As a result, plants and bacteria are slightly enriched in the carbon-12 isotope. This enrichment can be seen in present-day organisms and in carbon atoms that are contained in fossil organisms of all ages. Even the fossil stromatolites and bacteria from 3.5 b.y.a. show this enrichment, confirming that they are biological in origin (Figure 4.9).

The few sedimentary rocks that are older than 3.5 b.y., however, have been reworked too much to show any remaining fossils. They are metamorphic rocks, formed from sediments initially but then processed by pressure and heat until the original fabric was destroyed. Despite this, however, they retain any carbon isotopic signature left over from organic debris within them. And, in fact, some of these rocks show a carbon-12 enrichment that is identical to that seen in biota and in younger fossils.

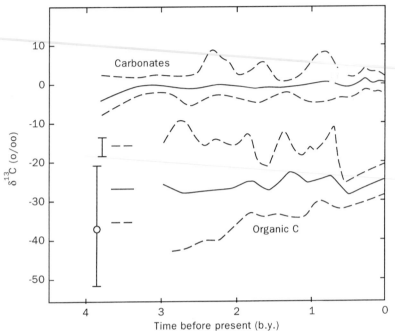

[Figure 4.9]
The history of terrestrial carbon isotopes. Values are in 'delta' notation, the parts-per-thousand deviation of the ratio $^{13}C/^{12}C$ from the ratio of a standard material. Values are shown both for inorganic carbonate rocks and for 'organic' carbon. In each case, the solid line represents an approximate running mean value and the dashed lines represent the extreme high and low values at each epoch. The 'organic' points showing substantial fractionation at 3.7 and 3.85 b.y.a. are interpreted as evidence that life may have existed at those times.

The oldest rocks that show this carbon isotopic signature were formed no later than 3.85 b.y.a. Because they have this signature, these rocks are thought to contain organic carbon, presumably a remnant of biological activity. This argument is not unique, however, because non-biological chemical reactions conceivably could produce a similar enrichment in carbon-12; it is suggestive, though.

Concluding comments

We are left with intriguing, but not unique, evidence suggesting that life existed as much as 3.85 b.y.a. Certainly, the fossils that have been discovered and dated as 3.5 b.y. old provide absolutely unique and convincing evidence supporting the existence of life at that time.

There are two additional important points that this evidence tells us about the origin and early history of life. First, a very short time had passed from the period when impacts would have continually sterilized the Earth and kept any life from gaining a foothold until the time at which life clearly existed. This time period, during which life must have originated on the

Earth, probably was no longer than about five or six hundred million years. It might have been as short as one hundred million years or less. Clearly, life originated relatively quickly once the Earth became inhabitable.

Second, it did not take life very long from the time of its formation until it had become widespread over the Earth and consisted of relatively sophisticated entities. The bacteria that were present 3.5 b.y.a. had very sophisticated chemistry, apparently based on DNA molecules to control reproduction, and some may have been photosynthetic. These species were not barely hanging on to life by a thin thread, but were robust life that was expanding into the available ecological niches.

72 If we are to understand the processes that led to the origin of life on Earth, and how life evolved to the high level of complexity that even the oldest and the simplest organisms seem to show, we must first understand a little about how life actually works. In particular, we would like to look at the ways in which life obtains energy to function, and to see how life functions in what seem to be harsh environments on the Earth. These will serve as guides to some of the possible constraints for life on the early Earth. As has become clear over the last several chapters, the environment on the early Earth was an extremely harsh one, at least by modern standards.

We will begin with a brief discussion of the chemical pathways that currently 'power' life. The diversity of pathways will be important as we begin to consider the possibility of life on other planets. Following that, we will look at organisms in some of the different harsh environments on the Earth and the sources of energy that they use. This will help determine what the limits of life might be. Some of what we might consider to be the harshest locales include hydrothermal vents near volcanic regions, where water is found at temperatures up to almost 400 °C; kilometers beneath the surface of the Earth; and frozen within rocks in Antarctica. Finally, the molecular basis for the genetic relationships between different life forms on the Earth will be discussed, including the structure of DNA and RNA; this genetic information indicates that life that is able to function in harsh environments probably dates back to the earliest history of life.

The chemical energy that powers life

We begin with a discussion of the chemical processes that are responsible for supplying energy to living organisms. This use of energy needs to be understood at the molecular level. Whether energy is obtained from the Sun through photosynthesis or from metabolism of existing molecules, the energy is utilized by chemical reactions. When two molecules react, they either give off energy in the reaction or they require energy to make the reaction go forward. The building of new cellular material, for example, requires energy to carry out the construction; the breakdown of existing molecules often gives off energy that can go toward this growth. We wish to understand how this energy can be tapped in order to perform useful functions.

In understanding how energy is utilized by living cells, we need to speak of the proper currency. The medium of exchange for energy in terrestrial life is the complex molecule adenosine triphosphate, known by its

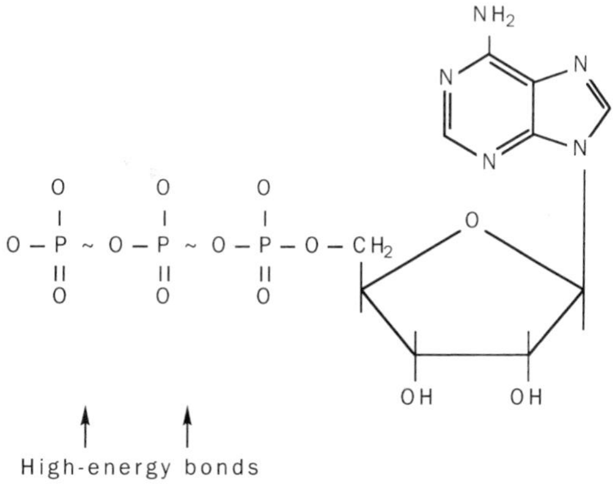

[Figure 5.1]
Structure of the ATP molecule.
Arrows point to the locations of the
two high-energy phosphorous
bonds.

abbreviation ATP. ATP contains the element phosphorous in groups that are bound with an oxygen atom; two of the three bonds between phosphorous and oxygen require the input of substantial amounts of energy to form (Figure 5.1). When chemical energy is needed for other purposes, these bonds can be broken, releasing that energy. These bonds can be created, and new molecules of ATP constructed, by chemical reactions that require an input of energy to proceed. In this manner, energy can be stored within a cell in the form of ATP, and can be accessed on demand as needed.

All living cells on Earth use the same currency in the form of ATP for transferring energy. This tells us something fundamental about the relationship between different species: since they use the same currency, either they must be descendants of the same species and, hence, of a single origin for life, or else multiple origins of life must have resulted in the same biochemical solution to the problem of storing energy. The second possibility seems unlikely unless ATP represents the only possible solution that might exist.

The energy required to construct ATP comes from chemical reactions between simple molecules. For example, hydrogen and oxygen can react to form water, giving off a finite amount of energy. The heat released is evident when the reaction is triggered, for instance by lighting a match in a room filled with hydrogen and oxygen. But how is this energy utilized? Let's look at a simple example. At the molecular level, we can consider the reaction between hydrogen and oxygen to form water as two separate reactions, each one involving the reactions between an element and an electron:

$$H_2 \rightarrow 2H^+ + 2e^- \tag{1}$$

and

$$1/2\,O_2 + 2e^- + 2H^+ \rightarrow H_2O. \tag{2}$$

These two reactions are referred to as 'half reactions', and combine together to give the overall reaction

$$H_2 + 1/2\,O_2 \rightarrow H_2O. \tag{3}$$

In this case, the first half-reaction involves the breaking down of a hydrogen molecule into two hydrogen ions and two electrons; the hydrogen is said to become oxidized. The second half-reaction involves the combination of these species with oxygen, and the oxygen is said to become reduced. The combination of reactions, one a reduction and the other an oxidation, is referred to as a 'redox' reaction.

In reducing a molecule or atom, what is being 'reduced' is the charge that it has. In this example, the oxygen takes on an electron, which has a negative charge, and its overall charge is reduced. The hydrogen gives up an electron in combining with oxygen, and its charge increases. Oxygen is so efficient at picking up an additional electron, and it is so prevalent in the terrestrial environment, that the effect that it has in chemical reactions is referred to as 'oxidation' and species that give up an electron become 'oxidized'. Because it gives up an electron in combining with oxygen, H_2 also is referred to as the electron donor. Similarly, because it accepts the electron, the O_2 is referred to as the electron acceptor. A redox reaction, then, requires both an electron donor and an electron acceptor, and it is the transfer of an electron from one molecule or atom to another in a chemical reaction that gives off energy that can be utilized to drive other chemical reactions. In other words, to track the energy in a chemical reaction, track the electron.

In biological systems, this type of reaction can occur with any molecule or atom that can act as an electron donor and any one that can act as an electron acceptor. The process of combining such molecules gives off energy. A wide variety of different chemical species can act as electron donors, including H_2, CO, Fe, Mn, CH_4, S, NO_2, H_2S, and NH_3 (along with virtually any organic carbon molecule). Similarly, a wide variety can act as electron acceptors, including O_2, CO_2, CO, S, NO_2, Mn, and Fe. Notice that the electron donors all can combine with oxygen and become oxidized, and that the electron acceptors all are oxidized species that can combine with hydrogen and become reduced. The metals, such as iron and manganese, can be present within various minerals in the environment.

Notice that some species can act as either an electron donor or an electron acceptor, depending on the circumstances (Figure 5.2). For example,

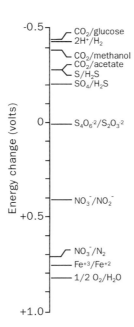

[Figure 5.2]
The energy 'ladder' of redox reactions. The energy required to drive a redox reaction is shown (for example, 0.43 volts to reduce the carbon in CO_2 to glucose); in each case, the oxidized species is shown on the left. Electrons donated from one reaction can be accepted in reactions lower down on the ladder, thereby providing a coupled redox reaction.

the CO molecule is intermediate in energy and can act as a starting product for producing a molecule with lower energy or as an end product starting from a molecule with higher energy. CO_2 can react with hydrogen to form CO and water, and CO can react further with hydrogen to form C or CH_4 and water.

Iron and manganese can be in the form of inorganic minerals that interact with biota. For example, Fe will have a +2 charge in FeO and a +3 charge in Fe_2O_3. As it makes the transition from one to the other, it becomes oxidized and gives off energy that can be utilized by some organisms.

The fact that there are so many different molecules that can act as electron donors or acceptors opens up the possibility that any of a wide variety of different chemical systems could power life. As we will see, in fact, different life forms on Earth can utilize many of these systems, depending on what chemicals are available in a specific location. We can see how these chemical reactions work in practice by examining a couple of specific systems. We'll start with examples of the most common types of biochemical pathways – photosynthesis, oxidation, and fermentation.

Photosynthesis may be the most familiar means by which energy is transported into biological systems. It is the process that takes energy from sunlight and uses it to drive useful chemistry. Because biota using photosynthesis get all of their carbon from inorganic CO_2 gas, they are called autotrophs (literally, 'self-feeding'; see Figure 5.3). They require only an

	Prefix
Carbon source:	
organic carbon	"hetero"
CO_2	"auto"
Energy source:	
organic carbon	"organo"
inorganic electron donor	"litho"
sunlight	"photo"

[Figure 5.3]
Definitions of the various prefixes used to name organisms, based on the organisms' sources of carbon and of energy.

energy source (sunlight) and a ready supply of chemicals in order to survive.

The photosynthesis process can be summarized in a simplified manner by the chemical reaction

$$CO_2 + H_2O + sunlight \rightarrow CH_2O + O_2. \qquad (4)$$

In this reaction, carbon dioxide and water are combined into organic material, with the energy of sunlight driving the chemical reactions. The two species on the right are CH_2O – which is used to represent any organically produced carbon (such as new cellular structure or organic material interior to the cell walls) – and O_2; the O_2 is given off as a gas, back into the atmosphere.

Clearly, this one reaction is a shorthand method of writing down a long series of reactions involving many different steps. Part of the series of reactions involves the ability of chlorophyll within plants to take the energy from sunlight and use it to break chemical bonds. Chlorophyll is the chemical that gives plants their color; the color comes from its ability to absorb some wavelengths of sunlight and not others. The absorbed energy is used to strip apart the hydrogen and oxygen in a water molecule; this is the reverse of the reaction described earlier as an example of a redox reaction, and here it requires the input of energy from sunlight to move forward. The protons (H^+) produced by the breakup of water provide the energy source for driving the production of ATP; the ATP then can be used to drive chemical reactions within the organism. In essence, the reducing power of hydrogen is used to reduce the carbon in CO_2 gas to a less-oxidized form, in this case referred to as CH_2O, and to convert it from a gas to a dissolved form of carbon that can be used by the organism for energy or for construction.

Oxidation is to some extent the reverse series of chemical reactions. Oxygen gas in the environment is taken up by an organism, where it combines with reduced molecules to release energy. A representative chemical reaction for aerobic respiration (utilizing oxygen from the atmosphere) can be written as

$$CH_2O + O_2 \rightarrow CO_2 + H_2O. \tag{5}$$

The oxidation of the carbon by its combining with oxygen releases energy that can be used to drive the production of ATP; again, the chemical energy stored in ATP can be used to power a wide variety of cellular activities. In this process, organic material is ingested by an organism (for instance, animals eating plants or bacteria eating free-standing organic molecules), and the molecules are oxidized to obtain energy.

In the absence of available oxygen to act as an electron acceptor, fermentation and anaerobic respiration are mechanisms by which an organism still can utilize organic carbon to drive energy production. These processes can be used by organisms that live in environments where there is no free oxygen available. An example of fermentation would be the reaction in which glucose, a sugar, produces ethanol and carbon dioxide:

$$C_6H_{12}O_6 \rightarrow 2C_2H_6O + 2CO_2. \tag{6}$$

This net reaction is not as efficient as oxidation, because only a fraction of the organic carbon in the glucose is oxidized completely to CO_2. That is, of the six carbon atoms in the original glucose molecule, two are oxidized to CO_2 and four remain in reduced form in ethanol. As a result, less energy is available by fermentation than by oxidation, where all of the organic carbon is oxidized to CO_2. The carbon in ethanol could be oxidized all the way to CO_2, but not through fermentation. In the case of fermentation, organic carbon is used both as an oxidant and a reductant, with the byproducts being carbon in very oxidized form (CO_2) and in a more reduced form (ethanol).

Species that use oxidation and fermentation get their carbon from previously produced organic molecules; they are called heterotrophs (from 'hetero' meaning others and 'troph' meaning to feed, they feed off of others; see Figure 5.3). Animals fall into this category, since they ingest organic carbon that was created by other organisms, and then get their energy by oxidizing the carbon.

Chemistry and life in unique environments

We can begin to look at other types of environments where organisms can exist, away from the 'standard' locations. For example, there are species that can get their energy by the oxidation of inorganic molecules, such as the bacteria that obtain their energy by oxidizing iron. In this case, iron serves as the electron donor and oxygen as the electron acceptor; the iron becomes oxidized and the oxygen reduced. The chemical reaction that describes this process can be written as

$$2Fe^{+2}+1/2\,O_2+2H^+ \rightarrow 2Fe^{+3}+H_2O. \tag{7}$$

Fe^{+3} and Fe^{+2} describe chemical states of iron, as discussed above. Bacteria that could oxidize iron may have been important early in the Earth's history, when oxygen was not present in great abundance, and may have participated in the production of the banded iron formations (see Chapter 4). The oxidation of iron proceeds so rapidly under neutral conditions that bacteria cannot compete effectively using this reaction. In very acidic conditions, however, the reaction proceeds much more slowly, so bacteria can make use it. As a result, iron-oxidizing bacteria thrive under very acidic conditions, such as in mine tailings. An example of such an iron-oxidizing bacteria is known as *Thiobacillus ferrooxidans.*

An intriguing example of bacteria whose energy consumption is related indirectly to the oxidation of iron is the discovery in 1995 of bacteria living deep beneath the surface, within the Columbia River Basalts. These rock formations are comprised of basaltic volcanic rocks which erupted to the surface in Oregon and Washington between 6 and 17 million years ago. They cover an area greater than 163 000 km^2 to a depth of 3–5 km. The bacteria use hydrogen as the electron donor. The source of the hydrogen is the chemical interaction between groundwater and iron in the basalt – the oxidation of the iron in basalt drives the production of hydrogen gas. The hydrogen is used to convert dissolved inorganic carbon into more-reduced forms and even into methane, which then can be found in substantial abundance in the groundwater. An example of a reaction utilized by bacteria that oxidize hydrogen is

$$6H_2+2O_2+CO_2 \rightarrow CH_2O+5H_2O. \tag{8}$$

Although the source of energy is the hydrogen, the source of the hydrogen is the oxidation of iron-containing minerals in the basalt by interaction with water.

The bacteria in the Columbia River Basalts are present more than a kilometer beneath the surface. They can be cultured in similar environments in the laboratory, and the oxidation of basalt to produce hydrogen also has been demonstrated in the laboratory. The result appears to be an ecosystem that is autotrophic, requiring no previously existing organic source of carbon, yet one that also does not rely on photosynthesis as a

source of energy. That is, this ecosystem exists entirely independently from
the surface biota. Of course, this does not mean that the bacteria originated in the deep subsurface. Rather, it seems more likely that, at some
point in the past, bacteria migrated into the deep layer of basalt through
groundwater and then became isolated from the surface.

There are a wide variety of redox couples that can provide energy for
biota. These include many different electron donors and electron acceptors. For example, sulfate-reducing bacteria utilize the energy from the
reaction

$$H_2SO_4 + 4H_2 \rightarrow H_2S + 4H_2O. \tag{9}$$

Methanogenic bacteria use one of the two reactions

$$CO_2 + 4H_2 \rightarrow CH_4 + 2H_2O \tag{10}$$

and

$$4CO + 2H_2O \rightarrow CH_4 + 3CO_2 \tag{11}$$

and acetogenic bacteria use the reaction

$$2CO_2 + 4H_2 \rightarrow CH_3COOH + 2H_2O. \tag{12}$$

In addition to iron-oxidizing bacteria there are also iron-reducing bacteria. They utilize the reaction

$$2Fe^{+3} + H_2 \rightarrow 2Fe^{+2} + 2H^+. \tag{13}$$

In this case, hydrogen is the electron donor, and the reduction of iron is
used to drive other chemical reactions.

There is an interesting class of hydrogen-oxidizing bacteria that thrive
and, indeed, can only survive in relatively hot environments. These species
are called 'hyperthermophilic', where 'thermophilic' refers to their love of
warm water, and 'hyper' means that they like extremely warm water.
Hyperthermophiles live in water warmer than 80 °C. The most extreme
example that has been cultured in the laboratory can thrive at temperatures up to about 125 °C. At a temperature this high, water would boil at the
surface of the Earth; therefore, these species can exist only in places where
high pressure keeps the water from boiling – for example, in hot springs at
the bottom of the ocean, where the weight of the overlying water keeps the
water from boiling, or deep beneath the surface of the Earth.

Hyperthermophiles, in fact, are notable in two specific locations – at
mid-ocean spreading centers at the bottom of the oceans and in hot
springs on the continents. Both places are associated with recent or active
volcanism and with the circulation of water through the hot subsurface

ocean water

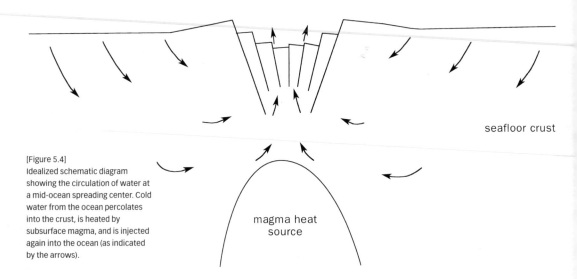

seafloor crust

[Figure 5.4]
Idealized schematic diagram showing the circulation of water at a mid-ocean spreading center. Cold water from the ocean percolates into the crust, is heated by subsurface magma, and is injected again into the ocean (as indicated by the arrows).

magma heat source

rocks and back to the surface (see Chapter 3). At mid-ocean ridges, the circulation of seawater occurs rapidly and can cycle the equivalent of all of the Earth's oceans through the vents in a very short time. At hot springs, the water comes from rainfall and percolates into the subsurface where it is heated and rises back to the surface.

Recent exploration of mid-ocean spreading centers by robotic submarines has shown a remarkable presence of life. Not only is there abundant bacterial life associated with vents of heated water, but substantial visible, macroscopic life as well. The water from the oceans is heated as it circulates beneath the ocean floor, and is injected back into the oceans at temperatures of up to 380 °C (at the high pressures at the bottom of the ocean, water will not boil until about 450 °C; see Figure 5.4). Along with the water comes abundant dissolved minerals and chemicals that precipitate as the hot water mixes with surrounding cooler ocean water; the water appears black as a result, and the plumes of hot water streaming back into the oceans are known as 'black smokers' (Figure 5.5). Some molecules will not precipitate out, including H_2, H_2S, CO, and Mn^{+2}, for example. These chemicals are in reduced form due to the chemical reactions that favor their production in the hot water as it mixes with the cooler ocean water; they are therefore able to serve as electron donors and provide energy that can power life.

Bacteria that are present live within the plume as it mixes with the cold

[Figure 5.5]
Mineral-rich hot water being injected into the ocean at a mid-ocean spreading center hydrothermal vent. The fuzzy cloud contains hot water (at temperatures up to 400 °C) with minerals precipitating into solid form as they enter the cooler ocean. These minerals build up a 'chimney' around the vent over time. (Photo courtesy of J. Childress.)

ocean. Although they cannot survive at the highest temperatures of the vent water, they appear to be able to survive in temperatures up to more than 125 °C. These bacteria get their carbon from the surrounding seawater; its ultimate source, though, is atmospheric CO_2 that has dissolved in the oceans. Because they get their CO_2 from inorganic sources and do not use photosynthesis as a source of energy, they are referred to as chemoautotrophs (Figure 5.3).

An example of continental hot springs with abundant hyperthermophilic life can be seen at Yellowstone National Park in northwestern Wyoming. There, a plume of hot material from the mantle has driven volcanic eruptions over the last several tens of millions of years (see Chapter 3). The site of the eruptions has been migrating to the northeast, and the most recent eruptions occurred about two million years ago. The volcanic

[Figure 5.6]
Hot spring at Yellowstone National Park. The spring is filled with water up to the brim, and the different shades are different colors of bacteria growing in the hot water. The water temperature at the opening, a couple of feet below the surface, is at about the boiling point. The pool of water is about 2.5 m across at the surface.

caldera has not erupted since then, but a reservoir of hot, molten rock sits only a few kilometers below the surface. Water at the surface percolates into the crust, where it comes into contact with hot rock. The heated water rises up to the surface, where it erupts as geysers and hot springs. As it reaches the surface, water is often at or near the boiling point.

Bacterial life is quite abundant in the Yellowstone hot springs. It shows itself as the multiple colors – vivid blues, yellows, and greens – associated with the pools of hot water (Figure 5.6). These colors are actually produced by colonies of bacteria that thrive in the hot water. The different colors are made up of species that thrive at different temperatures. The water closest to the center of the vent is the hottest, and it cools off as it moves farther out. Because different bacteria thrive at different temperatures, they end up residing at different distances from the vent and produce a 'bull's eye' pattern of colors. These bacteria are photosynthetic, and are capable of

[Figure 5.7]
Dry valley in Antarctica. Although there is almost no precipitation in the valley itself, runoff from the surrounding regions produces the ice-covered lake seen in the middle. (Photo courtesy of C.P. McKay.)

producing green chlorophyll to absorb the sunlight; when sunlight is abundant, however, not as much chlorophyll is needed, and the green color can be masked by other pigments in chlorophyll that can be bright orange, yellow or red.

Although most of the above discussion deals with microbial life that can use different sources of energy, there is an additional place that has an environment that would appear to be too harsh for life to survive – the dry valleys in Antarctica (Figure 5.7). The Antarctic region is an exceedingly harsh one. Temperatures typically are below the freezing point of ice and may rise above freezing for only a few days each year. In the dry valleys, the surrounding mountain ranges drive weather patterns that remove water from the atmosphere; as a result, the dry valleys themselves are deserts, with little or no precipitation. Water is very scarce, whether it be liquid or ice. Despite this, life is present.

In particular, there are bacteria that have taken up residence inside of rocks (Figure 5.8). This is not as difficult as it sounds. Rocks are very good at absorbing whatever sunlight might be available; the temperature of the rock can rise above the freezing point of water even if the air temperature is below freezing. The other requirements for biological activity might be physical access to water and to sunlight. Both of these will be available

84 within a sedimentary rock, which is composed of individual grains of rock less than a millimeter across, all packed and cemented together. The cement that binds the grains together does not completely fill the spaces between grains, and there will still be pore spaces that can hold water. The water can percolate into the rock on the rare occasions when it is available, and the small size of the pores will keep it from diffusing out during other times. Also, many of the individual mineral grains are sufficiently transparent for sunlight to be able to filter into the rock a short distance, perhaps to a depth of several millimeters.

The bacteria that exist within the rock are called lithoautotrophs – 'autotrophs' because they utilize CO_2 from the atmosphere, and 'litho' because they live within what appears to be solid rock (Figure 5.3). (They also can be called 'endoliths', since they reside inside rocks.) Yet again, we see that life has found a way to thrive by accessing the minimal resources that are available in an exceedingly harsh environment.

One of the common factors in all of the environments where life exists, and in all of the chemical reactions that power it, is the fundamental role of water. First, water acts as a solvent in which the different chemicals can dissolve. In this respect, water is often referred to as a universal solvent – because of its molecular structure, it can dissolve a wider variety of molecules than can most other liquids (see Chapter 7). Second, and perhaps more importantly, water acts as a medium in which chemicals can move – it provides a place for chemical reactions to occur and a medium of transport by which the reacting species can be brought together and the reaction products can be taken away. If these reactions were to take place in air, for example, it would be much more difficult to manage the different chemicals. On Earth, the presence of liquid water is a key to the existence of life.

[Figure 5.8]
A slice through an Antarctic sandstone. The individual grains of sand comprising the rock are visible. The layers that are visible contain bacteria that live within the rock, despite average temperatures well below the freezing point of water. The image is 1.8 cm across. (Photo courtesy of E.I. Friedmann.)

What is a 'standard' and what is a 'harsh' environment? 85
The 'tree of life'

The above discussion of life in harsh environments probably leaves the impression that the life with which we are most familiar – that which is all around us at the surface of the Earth – is the 'standard' form of life against which all other life should be compared. In this context, then, life in hot springs, in Antarctica, or in saline or acid environments, seems somewhat odd and out of place. There is a line of evidence, however, that indicates exactly the opposite. It may be hyperthermophilic life that is the standard, and everything else may have evolved to be different from it. To explore this possibility, we first need to discuss the biochemistry of DNA and RNA.

DNA is the abbreviation for deoxyribonucleic acid. It is the molecule in living species that contains the hereditary information necessary to allow an organism to replicate itself. A DNA molecule consists of a long double-helix structure of molecules (Figure 5.9). A single helix is described, for example, by the outside of the threads on a screw; they wrap into a corkscrew-like structure. In a double helix, two of these are intertwined together and are connected to each other by molecular structures. The 'backbone' of each helix consists of a series of molecules composed of the elements carbon, hydrogen, oxygen, and phosphorous, bonded together to form specific structures (the 'deoxyribose' in the DNA name). The two

[Figure 5.9]
Structure of the DNA molecule. The double helix consists of alternating sugar and phosphate structures, and the 'cross bars' consist of pairs of base structures. Ten base pairs are contained within one complete loop of the double helix, and one loop is 3.4 nm long.

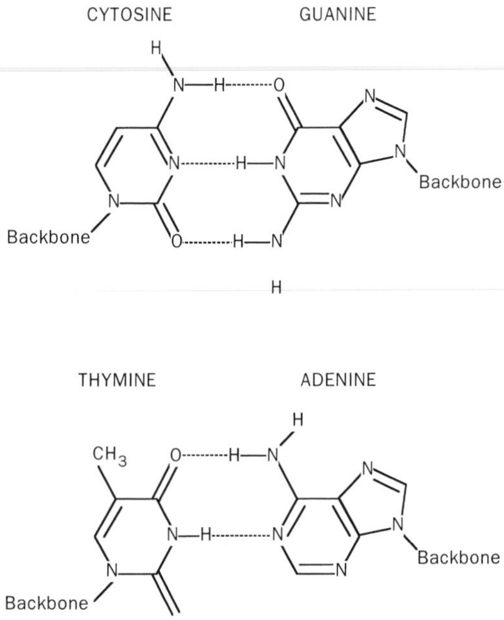

CYTOSINE   GUANINE

THYMINE   ADENINE

[Figure 5.10]
Structure of the four bases in the DNA molecule. The dashed lines indicate where the two complementary bases bond together. The deoxiribose 'backbone' of the DNA molecule is not shown specifically, but its location is indicated.

strands forming the backbones are connected by structures of molecules that are termed 'bases'.

There are four bases, each containing a double ring of molecules. The four bases in DNA are cytosine, thymine, adenine, and guanine; these are usually referred to by their first letters, C, T, A, and G. The four molecules differ slightly in their composition, and they are all slightly different sizes. However, the bases can bond chemically to each other and to the two back-bones of the DNA. When they combine, G and C can bond together and A and T can bond together (Figure 5.10). The combined G and C structure is the same size as the combined A and T structure, and each one will fit exactly between the helical backbones of the DNA molecules. One can envision the overall structure by thinking of a ladder that has been twisted so that the uprights of the ladder form a double helix (Figure 5.9). The double-base structures form the rungs of the ladder, connecting the helical strands. Because the two pairs of bases are the same size, and each can bond onto the backbone, the ordering of bases is variable. It is the ordering, in fact, that contains the genetic information.

This structure immediately suggests how the DNA molecule can perform its two important functions of making an exact copy of itself and of utilizing the genetic information. If the two helixes that compose the double-helix structure split in two down the middle between the two bases connecting them, as one might unzip a zipper, then each individual strand can act as a template for the construction of a complete DNA molecule that would be identical to the original. Starting with one strand that consists of a backbone and the bases that are attached to it, structures that contain a base and a part of the backbone can attach themselves to it to create a new double-helix. When they attach, they do so only in the way that fits together physically – a G base will bond to a C base, and an A base to a T base. If the wrong base-containing structure tries to bind onto the first helix, it won't fit into the available space. The end result is that each strand, each half of the zipper, can turn itself into a complete DNA molecule that is a duplicate of the original; the one DNA molecule can become two identical and complete DNA molecules.

The genetic information in a DNA molecule is contained in the sequence of bases along the chain, much as Morse code consists of a pattern of dots and dashes that contain information on the letters of the alphabet. The information contained in the DNA molecule drives the operations of the cell and determines its functions and behavior. The method by which this is accomplished is through interactions with RNA.

RNA, ribonucleic acid, is a companion molecule to DNA. RNA generally is single stranded, rather than double-stranded like DNA, and it uses the base uracil (U) rather than thymine (T); the composition of U is slightly different from that of T, but it has the same overall size. The backbone of an RNA molecule has a slightly different composition from that of a DNA molecule – it differs only in whether an extra oxygen molecule in the structure of the backbone is present (RNA contains ribose instead of deoxyribose). RNA can be constructed from DNA by a process similar to the duplication of DNA. A DNA molecule can split in two, and an RNA molecule can be constructed on the template of the single strand of DNA.

There are several different types of RNA, depending on their function in a cell. Messenger RNA (mRNA), transfer RNA (tRNA) and ribosomal RNA (rRNA) perform different tasks relating to the incorporation of amino acids into longer structures and the construction of proteins from amino acids. The DNA molecule is copied or 'transcribed' into a molecule of mRNA, which then contains the same genetic information as the original DNA molecule. The information coded on the mRNA then is read or translated by tRNA using entities within the cell called ribosomes, and enables the construction of a new protein; the ribosomes themselves are composed of a combination of ribosomal RNA (rRNA) and proteins

Clearly, the nature and structure of DNA and RNA are important for understanding how cells multiply, how genetic information is passed on and, through mistakes or mutations in copying the strands, how evolution

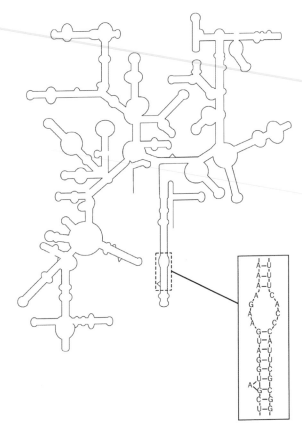

[Figure 5.11]
Structure of the 16S ribosomal RNA molecule from an *E. coli* bacterium. The inset shows some of the individual bases in the molecule and how they are strung together, while the main figure shows the complicated shape that results. (After Brock *et al.*, 1994.)

might occur. Here, however, we are interested particularly in using the genetic information to gain insight into the evolution of species on Earth and the nature of the oldest common ancestors of terrestrial life. Certain parts of the genetic sequence can be used as a measuring stick to show us how far different species have evolved from each other.

Specifically, let's examine the sequence of bases in a particular rRNA molecule. There are several different rRNA molecules in cells, having different numbers of bases in the strand. The (16S)rRNA molecule in pro-karyotes (and the almost identical (18S)rRNA molecule found in eukary-otes) contains about 1500 bases within its entire length (Figure 5.11). As a cell constructs new (16S)rRNA molecules, they should be identical copies of the existing ones. As long periods of time pass, however, errors will creep into the copying process, and the sequence of bases will begin to change.

All species contain the same types of rRNA molecules, a remnant of the derivation from a single common ancestor. Although all species came from a common ancestor, they have diverged away from each other by these random mutations. However, the need to produce proteins accurately, with errors often being lethal, means that changes by mutation or evolution will occur slowly. As errors creep into the sequence of bases, though, they will get passed on to the offspring of that species. And, there will be areas within the (16S)rRNA molecule that will be almost identical among all species, that will have almost the same sequence of bases. Thus, we can look at the sequence of bases in the (16S)rRNA molecule in different species, and use the number of differences as an indicator of the sequence in which various species diverged from their common ancestor. In this comparison, a sequence of bases that is not actively used by the cell must be examined; if an active sequence were used, then natural selection may influence the mutations that produce changes in the sequence. Fortunately, there are many sequences of bases that are not used actively in cellular functions.

For two species to have almost identical (16S)rRNA sequences, they would have to have diverged from each other very recently; two species with very different sequences, on the other hand, must have diverged a longer time ago. It turns out, for example, that all animals are very close to one another in terms of their genetic distance based on (16S)rRNA. All bacteria are relatively close to one another as well, but animals and bacteria are relatively far from each other. This makes sense, because we imagine that animals and bacteria are so different that they must have diverged a very long time ago, but animals are sufficiently similar to each other that they would have diverged more recently. The information contained within the geological record (Chapter 4) leads us to the same conclusions, with plants and animals appearing relatively recently.

The sequence of bases for a large number of species can be compared to each other to get a sense of how far apart all of the different species are. By this process, a 'tree' can be constructed that shows the evolutionary positions of the species (Figure 5.12). Since this 'tree of life' is based on the genetic differences between species that all are alive today, it does not place the species into an historical sequence, but it can be used to gauge the genetic distances between species. This allows us to make an estimate as to the relative timing at which species diverged evolutionarily from their common ancestors. We would need to know the rate at which mutations occurred in the copying of rRNA in order to estimate the timescale for evolution; in fact, the rate has not necessarily been constant throughout time.

The tree of life divides naturally into three branches, now termed 'domains' (Figure 5.12). These domains are the eukarya (cells containing

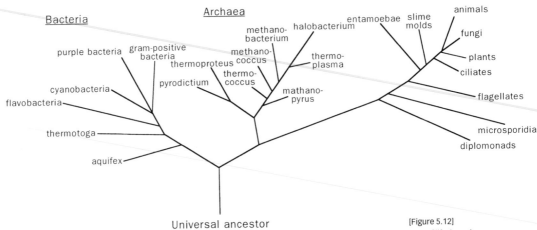

Bacteria

flavobacteria

cyanobacteria

purple bacteria   gram-positive
bacteria

thermotoga

aquifex

thermoproteus

pyrodictium

Archaea

methano-
bacterium

methano-
coccus

thermo-
coccus

halobacterium

thermo-
plasma

mathano-
pyrus

Eukarya

animals

entamoebae slime
molds

fungi

plants

ciliates

flagellates

microsporidia

diplomonads

Universal ancestor

[Figure 5.12]
Tree of life, based on gene
sequences in rRNA molecules.
The line segments represent
the genetic distance that each
organism has evolved since
splitting off from the rest of
the tree. The three domains
of life are indicated.

nuclei, and including plants and animals, for example), the bacteria, and a third, newly recognized grouping, called the archaea. The archaea differ from bacteria in such basic properties as the structure of their cell walls; they clearly represent a fundamentally different type of organism. These domains as defined by molecular genetics are completely different from the five 'kingdoms' of life that were thought to exist 30 years ago (plants, animals, fungi, bacteria, and protists, or single-celled eukaryotes). These kingdoms were based on the physical differences between species rather than on their genetic differences. Clearly, the concept of kingdoms as based on morphology and appearance is obsolete now that we have a way to determine the genetic connections between species directly.

While the tree of life can be constructed showing the differences between species, this approach does not allow a unique determination of the location of the 'root' of the tree – the position that represents the divergence of all presently living species from their last common ancestor. The reason is that there is no way to know what the sequence of genes was in the earliest organisms, which means that evolution could have proceeded in either direction. However, there is a second approach that can be used to find a unique location for the root. In a number of instances, there is a sequence of genes that has been duplicated. The duplicated sequence appears in species in all three domains, meaning that it must have been duplicated prior to the time of the last common ancestor. The presence or absence of these duplicated sequences provides a unique starting point, from which the sequence of divergence can be determined.

This technique does not point to a single species as being the common ancestor to all life, because every known species has evolved away from **91** this root by some amount, but it does show us which species have evolved least from that common ancestor and which domains are closest to it. The location of this root suggests that the first major evolutionary divergence from the last common ancestor was into bacteria and archaea, and that the eukarya diverged from the archaea somewhat later in history. There are no living species that are genetically very close to the location of this root. This implies that all life has evolved significantly since the first use of RNA in cells.

The significance that the 'tree of life' has for the occurrence of life in harsh places is that all of the species closest to the location of the root of the tree of life are hyperthermophilic in nature – they live and thrive in hot water. This does not necessarily require that life originated in hot springs, although this is a possibility (see Chapter 6). Rather, it suggests that all of the life on Earth today descended from species that did live in hot springs. If life did not originate in hot springs then it must have evolved through a stage at one time in which it lived in such an environment; in such a case, however, we would have no evidence regarding the nature of any life that existed prior to the hyperthermophilic life. We see that the norm for life on Earth may be hyperthermophilic!

Similarly, many of the species closest to the root also are methanogens, which utilize CO_2 as their electron acceptor. It may not be surprising that these organisms are very ancient, given the tremendous availability of CO_2 in the early terrestrial environment (see Chapter 4).

Concluding comments

We have seen that life on Earth enjoys a spectacular diversity. The energy that powers biological activity comes from an amazingly diverse set of sources. It would seem that almost wherever energy is available – in the form of sunlight powering photosynthesis or chemical energy from redox reactions – there exists organisms that can take advantage of it. The availability of chemical drivers for life is, itself, tremendously widespread. Almost everywhere on Earth, there exists chemical disequilibrium that can supply energy for life. It is almost as if life exists today everywhere that there is a chemical reaction able to provide energy for it. As a result, microbial life – the most widespread and abundant form of life on Earth – occupies a tremendous breadth of ecological niches. There are photo-autotrophs that get their energy from sunlight and their carbon from atmospheric CO_2. There are heterotrophs that obtain carbon from organic molecules created by other species. And there are chemolithoautotrophs that get their energy from inorganic chemical reactions within rocks. The common denominator appears to be an input of energy from the Sun or from reactions which cause a chemical system to move out of equilibrium, with the movement back toward equilibrium powering life.

There are two important conclusions we can draw from the previous discussion, dealing with life on Earth and the possibility of extraterrestrial life.

With regard to life on Earth, we see that energy is available from a wide variety of sources. When we consider the actual formation of life more than 3.5 billion years ago (in the next chapter), one of the major considerations will be the availability of energy to power biological activity. Although the ways in which present-day life accesses and utilizes energy are no doubt very complex, and must have evolved substantially since life's origins, it is exceedingly clear that there must have been a large number of available sources of energy. The first life forms could have tapped into any of them. And, in the absence of observations that tell us directly about the origin of life, we need to keep our minds open as to which is the most likely to have been used by early life.

With regard to extraterrestrial life – whether it be in our solar system or elsewhere – again it would appear that a wide variety of energy sources could be used effectively by biota. In looking for where extraterrestrial life could exist, then, we will use our basic knowledge of physics, chemistry, and geology to try to understand where the available sources of energy might be. The wide variety of locations on Earth that can sustain life suggest that it can exist in what used to be considered an extremely harsh environment. Ultimately, we can imagine that we will find that any location with access to the chemical reactions able to provide the energy to power life may be able to support life. The implication of this conclusion is immediately apparent – there is the possibility that life may be remarkably widespread throughout the universe!

Several of the previous chapters have centered on a discussion of the
environmental conditions on the early Earth and the nature of some of the
earliest life. These provide what can be thought of as boundary conditions
for the origin of life. They do not tell us directly about life's origin. Here, we
will discuss the processes by which life might have begun on the Earth.

There is a major conclusion that we can start out with, although it may
seem obvious – life did begin on the Earth. Even after we discuss the processes that potentially account for life arising out of an inert planet, it
might be tempting to conclude that they are just too improbable to have
occurred. Some people think that an origin of life is so improbable, or that
life on Earth appeared so quickly after it became possible, that life must
have been implanted from elsewhere. This concept of 'panspermia',
however, merely transfers life's origin to some other location; it does not
solve the problem of understanding the origin itself.

Another important conclusion is that, today, we do not understand the
origin of life. Certainly, we understand much about how life operates, the
conditions under which life originated, and the possible sequence of
events surrounding the origin of life. However, the events that took us from
prebiological chemistry on the early Earth to life that operates as we know
it, using DNA and RNA molecules to perform many of the genetic functions, are not understood. That said, however, we can describe some of the
processes that must have occurred and some that might have occurred,
and that will be our goal. Obviously, our inability to understand fully life's
origins on the Earth affects our ability to determine whether life can have
undergone a separate origin elsewhere, and whether it might exist elsewhere. This lack of understanding of some of the processes, however,
should not keep us from concluding that conditions on other planets may
have been similar to those on the Earth and that life may exist there.

We will begin with a discussion of what constitutes life, since we are
looking at understanding the boundary between absence and presence of
life. We then will discuss the prebiotic chemical processes that occurred
and the mechanisms by which life may have originated. Through this sort of
a discussion, we will be able to begin to define the conditions that we would
look for in our search for places other than the Earth where life might exist.

What is life?

In order to begin a discussion of the origin of life, we need to agree on what
constitutes life. At the extremes, we think that we can identify what is living

94 and what is not. Animals, for example, are living, and rocks are not. The most basic definition of life might involve the ability to ingest nutrients, to give off waste byproducts, and to grow and reproduce. The definition of what is a nutrient and what is a byproduct is difficult, since different species can utilize a wide variety of sources of energy. In fact, we saw in Chapter 5 that the waste byproduct of one species might be the nutrient for another.

We also have a problem with the concept of growth. Mountains grow, although on geological timescales. Individual mineral crystals grow. By freeze/thaw weathering, in fact, rocks or crystals can break, one can become two, and each can further grow. Yet we know intuitively that these are not alive.

Fire has been brought up as an example of something that might meet a simple definition for life but which we do not normally think of as living. Fire takes in nutrients (oxygen and combustible fuel) and gives off waste products (heat, carbon dioxide, and smoke). It grows, by expanding its geographical coverage to new, unconsumed areas. However, we generally think of fire as a byproduct of nonbiological chemical reactions – the oxidation of the fuel, in which heat is released, with the fire itself being the visible manifestation of this heat.

Another confusing example that has been mentioned is the mule. A mule is the offspring resulting from the breeding of a horse with a donkey; it is a sterile animal that cannot reproduce. Being incapable of reproduction, does this mean that it is not alive? The individual cells in a single mule do reproduce and metabolize. Are the components alive while the whole is not?

How about viruses? Although they contain DNA, they are not able to reproduce on their own. They propagate by invading cells, using the chemical machinery in that cell to produce a large number of copies of the original DNA, and then breaking apart the cell and sending the offspring out as new invaders. Viruses clearly are capable of reproducing, but they cannot do so on their own. They are on the boundaries of what may be considered as living.

Despite the obvious difficulties, we can make some general statements about what constitutes life. Living entities utilize energy from some source to drive chemical reactions, are generally capable of reproduction, and can undergo evolution. The concept of evolution is one that is central to the history of life on the Earth. Over geologic time, new species have been created by the modification of existing species. This process, first described by Charles Darwin in his book on the origin of species in 1859, involves inexact reproduction. If an organism can produce offspring that contain slight differences from the parent, then those offspring that have changes that are somehow beneficial will reproduce more efficiently than those without; they will be more successful at competing for resources and at propagating themselves. If the changes are passed on to subsequent generations, modified organisms will eventually outnumber and replace the original organisms.

Again, we have to be careful about including the ability to evolve as part of the definition. It cannot apply to a single organism, but rather to an entire species. And, what about species such as the single-celled bacteria that appear to have gone unchanged since the oldest known life some 3.5 billion years ago? They do not appear to have changed significantly during that time, although they are capable of it.

Finally, what about the example given by A.I. Oparin in his classic book on the origin of life published in 1936? He asks about an army of robots that are capable of creating new robots. Is this life? Or how about a computer program that allows 'species' to reproduce, either with or without modification. Is that life? These are philosophical questions that fall outside of the present discussion, but are, at the same time, key to understanding what life is.

We need to recognize that any definition of life will be based on what we know of living and nonliving entities that we come in contact with, and that it will be ambiguous. Any definition which tries to describe a unique set of characteristics of life is bound to fail. There will always be exceptions or counter-examples, and such a definition will always be determined by the suite of organisms that we see today on the Earth. Similarly, at the boundaries between the living and the nonliving, any single definition will fail. We need to keep this in mind as we discuss the origin of life. The boundary between nonliving and living entities in an evolutionary sequence probably will be just as confusing as the boundary is today.

Necessary prerequisites for life

For life to exist, it needs access to the basic building blocks of organic molecules. These consist of what is known as the biogenic elements, primarily carbon, hydrogen, oxygen, nitrogen, phosphorous, and sulfur. The central role played by carbon molecules is key. Carbon forms a wider variety of chemical bonds than any other element, and is present in all of the biological chemicals as a major player. Is life possible based on some element other than carbon? Although it cannot be ruled out, it is unlikely (see Chapter 7).

96 Access to the biogenic elements is not a terribly difficult constraint. As a natural consequence of the geological processes that occur on large, rocky planets like the Earth, Venus, or Mars, these elements are readily available at the surface and in the interiors of the solid planets.

The origin of life needed a source of organic molecules as building blocks for the more-complicated organic molecules such as the proteins that actually comprise living organisms. These building blocks might include simple molecules such as methane (CH_4) and ethane (C_2H_6), or more complicated molecules such as amino acids (which include chains of up to six carbon atoms bonded to other elements). Living entities would also need a mechanism by which they could reproduce. This generally is thought to require some means for passing on genetic information, and so would demand taking these simple molecules and constructing more complicated organic molecules. Although DNA molecules can be thousands of atoms long, such a complicated molecule can be presumed to have evolved from shorter chains over time; the original life, however, would need to contain molecules more complicated than the simplest organic molecules.

Finally, an origin of life also required a means to access and utilize energy. In modern life, the energy can be sunlight (through photosynthesis) or chemical (see Chapter 5). In the earliest life, it was most likely chemical, since the chemical machinery needed to utilize sunlight is relatively complex.

We will discuss each of these requirements in more detail, and then try to put them together into a unified view of the origin of life.

The occurrence of the first life probably required the existence of a ready source of organic molecules. The simplest life imaginable would be one that accessed existing organic molecules as a source of both energy and building materials. Organisms that can use the carbon in, for example, CO_2 to build up more-complex molecules would need to be much more complicated than if they could use any preexisting organic molecules directly. How might organic molecules form?

The first, and still most widely accepted, answer was proposed by Harold Urey and demonstrated by Stanley Miller in experiments performed in the 1950s. Approaching the problem of organic molecules from a chemical perspective, they recognized that complex molecules such as amino acids, the building blocks for life, consisted of carbon atoms that were intermediate between fully oxidized carbon (that would be present as CO_2) and reduced carbon (as CH_4 or, in the extreme, C). If they could instigate chemical reactions that combined oxidation of reduced carbon (such as methane) with reduction of some other species, they might produce organic molecules as a byproduct.

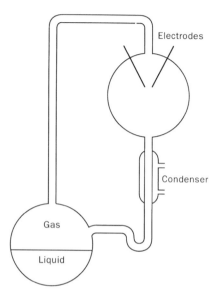

[Figure 6.1]
Schematic diagram of the apparatus used in the Miller–Urey experiment. Liquid water was heated in the presence of ammonia, methane, and hydrogen gas. The gas mixture was 'sparked' using electrodes, and the resulting chemicals were condensed and allowed back into the original chamber. On analysis, organic molecules were found in the liquid.

Miller created a laboratory experiment to simulate chemical reactions on the prebiotic Earth (Figure 6.1). In the 1950s, that environment was thought to be very reducing, consisting of molecules such as methane (CH_4) and ammonia (NH_3) in addition to water. A reducing environment seemed plausible because the gas out of which the Earth, planets, and Sun formed was very rich in hydrogen; with so much hydrogen available, more oxidizing species such as CO_2 or N_2 would combine readily with the hydrogen. In addition, the Jupiter atmosphere, also thought to be representative of the primordial gases, is very highly reducing – it consists primarily of hydrogen, with other constituents being present in small amounts.

The experiment consisted of a flask partly filled with liquid water and containing gaseous methane and ammonia. Water vapor was created by heating the liquid, and lightning as an energy source to trigger chemical reactions was simulated by an electric spark discharge in the gas. The products of any chemical reactions were recycled into the liquid water by condensation. After the experiment had run for several weeks, the contents of the flask were analyzed. They contained a wide variety of important organic molecules. These included hydroxy acids, aliphatic acids, urea, and, notably, amino acids. About a sixth of the carbon initially in the flask had turned into organic molecules; the amino acid glycine alone accounted for about 2% of the carbon atoms. The amino acids did not form directly from the initial gases, but formed by the reaction of hydrogen cyanide (HCN) with some of the CHO-containing compounds; these simpler organics were created directly from the initial gases.

Other sources of energy, such as solar ultraviolet light or cosmic rays, were found to have the same effect. However, the yield of organic molecules and amino acids was much lower when the atmosphere was not reducing. When the nitrogen was present as N_2 rather than NH_3, and the carbon as CO_2 rather than CH_4, the yield of organics was as much as one thousand times less. This is not too surprising given the difficulty of creating partly reduced molecules in an oxidizing atmosphere – any organics would react readily with the gases and produce the more oxidized forms of carbon.

If the early Earth had a reducing environment, then it is very plausible that the processes envisioned by Miller and Urey would act to create a rich soup of organic molecules in the oceans. Such a soup would act as fuel for the first living entity. One potential problem with this scenario, however, is that the atmospheric methane and ammonia, which are so important to the overall process, would not have been stable in the early terrestrial environment. They are subject to photodissociation by solar ultraviolet light, a process in which a molecule absorbs a photon of light and the energy that it receives is sufficient to break up the molecule. Over time, the methane and ammonia would have been broken apart, and the hydrogen would be able to escape from the atmosphere; this would result in the reducing atmosphere becoming more oxidized.

A second possible source of organic molecules on the prebiotic Earth was suggested recently, based on the discovery of widespread hydrothermal systems on the ocean floor (see Chapters 4 and 5). These locales are very rich ecological environments, supporting abundant biological activity on chemical energy. Importantly, though, such environments are also capable of producing organic molecules, and may have done so on the prebiotic Earth.

To understand the chemical pathways by which organic molecules are produced in hydrothermal systems, we need to follow the physical pathways of the water circulating through the crust. At mid-ocean spreading centers, basaltic magma is injected from the upper mantle into the crust. Water is able to circulate from the ocean into this crust, percolating through cracks created as the rock cools and moves away from the ridge. The water is heated up to as hot as 1000 °C, and is injected back into the ocean. Water entering the ocean has been measured as being as hot as 450 °C; the bulk of the water is not this hot, but reenters the ocean at somewhat lower temperatures (see Figures 5.4 and 5.5).

The water that is heated up to the highest temperatures will be able to

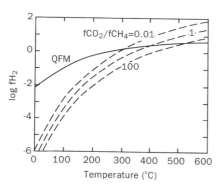

[Figure 6.2]
The oxidation state of gases as they come out of the mantle, as a function of temperature. fH_2 (the 'fugacity') is equivalent to the partial pressure of hydrogen, and is determined by chemical interactions with the minerals; the result is that fH_2 will follow the 'quartz/fayalite/magnetite' buffer, and will fall along the solid line labeled 'QFM'. The dashed lines indicate the ratio of carbon in CO_2 to that in CH_4 for specific values of fH_2 and temperature. Along the QFM buffer at high temperatures, most of the carbon will be in the form of CO_2 ($fCO_2/fCH_4 > 1$). At lower temperatures, however, the carbon will prefer to be present as CH_4. The chemical reactions that turn CO_2 into CH_4 may not go to completion, leaving some of the carbon in intermediate states as various organic molecules. (After Shock, 1992.)

equilibrate chemically with the nearby rocks. Because the rocks came from the upper mantle, their oxidation state will be similar to that of the mantle, which is relatively oxidizing. In the mantle, the oxidation state refers to how much oxygen has combined with the silicate minerals. Specifically, the amount of oxygen available for reaction is determined by the chemical equilibrium between interactions of several minerals. Possibly the most important equilibrium involves interaction between the minerals quartz, fayalite, and magnetite, the so-called 'qfm' buffer. Mantle carbon that is in equilibrium with these rocks will be predominantly CO_2 rather than CO or CH_4; nitrogen will be predominantly N_2. This is the case whether the source of the carbon and nitrogen is within the mantle itself, or is the addition of extraterrestrial carbon and nitrogen (to be discussed later).

As the water is injected back into the ocean after circulating through the crust, it mixes with the surrounding ocean water and cools off. At lower temperatures, the chemical interactions with the qfm minerals push the reactions in the other direction, so that some of the carbon will exist as methane (Figure 6.2). At temperatures below 500 °C, however, the reaction to form methane is inhibited from proceeding to completion. Rather, intermediate species will be formed involving carbon that is partially, but not fully, reduced. These will include organic molecules of the same type produced in the Miller–Urey experiments. Although these molecules are not stable, they can be long-lived transient species. In fact, once the organic molecules are produced, they will have the same lifetimes as organics produced by atmospheric reactions. The organic molecules will be able to accumulate in the oceans and provide a source of molecules for forming life.

The final potential source of organic molecules is related to incoming comets, asteroids, and meteorites. They can be a source of organic molecules, since they themselves contain organics. In addition, they can trigger the chemical production of organics in the extreme pressure and temperature environments associated with their passage through the atmosphere and impact onto the surface. The origin of life is closely tied to the rate of accumulation of material by the accreting Earth. Sufficiently large impacts can destroy the clement environment that would support life (Chapter 2), and it is only with the decrease in the rate of accumulation of material that life could exist.

It seems odd at first that organic molecules contained in incoming debris could withstand the high temperatures resulting from an impact event. However, the smaller particles can enter the upper atmosphere of the accreting Earth and can be slowed down gently enough that they are not heated a great deal. Organic molecules have been found directly in meteorites collected at the surface of the Earth, for example, and in interplanetary dust particles that have been collected from the stratosphere (Figure 6.3). These objects contain the same sorts of organic molecules that are produced in the Miller–Urey experiments, including molecules as complicated as amino acids. These molecules were produced in the interplanetary and interstellar medium during the earliest history of the solar system; they are not indicative of the existence of life in the interplanetary medium.

In addition, larger impacting objects can create organic molecules in the atmosphere due to the shock heating effects produced by their passage through the atmosphere, by 'airbursts' within the atmosphere, or by the supersonic expansion of the ejecta plume that results from the impact. The shock wave on entry is generated by supersonic travel through the atmosphere, as the impact velocities are much greater than the speed of sound and the air molecules cannot easily get out of the way. Objects as large as several hundred meters in size may undergo catastrophic airburst due to the pressure of the shock wave and are unable to make it to the ground

[Figure 6.3]
The carbonaceous chondrite meteorite Murray, which fell in Kentucky in 1950. The light colored inclusions are dominantly chondrules, and they are set in a dark, fine-grained matrix. This type of meteorite has a composition very similar to the Sun (except for H and He) and has been processed very little since the formation of the solar system. This stone is about 4 cm across. (Photo courtesy of A. Brearley.)

intact; when they burst, all of their kinetic energy is put into the atmosphere and triggers a shock wave. Finally, after a large object hits the ground, the ejecta tossed out from the impact site will push on the atmosphere and generate another shock wave. Laboratory experiments have demonstrated that shock waves can generate organic molecules such as HCN, simple hydrocarbons such as C_2H_2 and C_2H_4, and soot; amino acids have also been produced.

So, which of these processes – atmospheric formation triggered by lightning and ultraviolet light, formation in hydrothermal systems, or formation triggered by impact – would have been the dominant source of organic molecules on the prebiotic Earth? Clearly, the estimates of the net production rates of organics are uncertain. They depend on processes that are not well understood, and on the composition of the atmosphere which also is not well known. Despite this, we can make a preliminary estimate.

In a reducing atmosphere on the early Earth, production of organics by lightning might be as high as 3×10^9 kg/yr, production in hydrothermal systems 2×10^8 kg/yr, and production from atmospheric shock due to impact 2×10^{10} kg/yr. In an atmosphere that is not reducing but is more neutral, lightning production might be 3×10^7 kg/yr, hydrothermal system production 2×10^8 kg/yr (it does not depend on the oxidation state of the atmosphere), and impact production of only 400 kg/yr. By comparison, the supply of organics brought in by accreting material, primarily from interplanetary dust particles, would have been about 6×10^7 kg/yr.

Clearly, each potential source of organic molecules might have been expected to play a role in the total production of organics. We do not know enough about the oxidation state of the early Earth or about the details of the production of organics to decide which is the dominant factor. For the time being, we will have to keep in mind that there were multiple mechanisms by which organic molecules could have been produced, and that each one may have played a role in the overall production. However, it is clear that organic molecules would have been widespread on the early Earth, and that they would have been available for use in the origin of life.

The origin of life

How did we go from the presence of nonbiologically produced organic molecules, including molecules perhaps as complex as amino acids, to entities that are 'alive' in the generally understood sense of the term? The gap between nonliving chemicals and the simplest type of life today is incredibly large. Even the simplest life today, or the types of organisms that are inferred to have lived as long ago as 3.5 billion years ago, are based on DNA and RNA chemistry. They involve very complex chemical interactions and catalytic enzymes that help to transfer information from one molecule to another and to create complex proteins. Can life exist without this complex chemistry? Ultimately, this again leads us back to the question of what constitutes life. If we think of the requirements for life discussed earlier, we recognize its two basic functions of taking in energy and using it to grow and multiply. The function of reproduction involves the transferal of genetic information from one living entity to another; this is done today using DNA and RNA.

In reproduction, information is coded into a long strand of DNA by the sequence of what are called nucleotides (see Chapter 5). The DNA molecule is copied or transcribed into an RNA molecule that then contains the same genetic information. The RNA molecule is used to create protein molecules, using this information. The proteins ultimately are used to create new cells, and they also contribute to the functioning of the reproduction process by acting as catalysts in the formation of the RNA molecules.

We can imagine a simpler world than one in which both DNA and RNA are involved in the reproduction of cells. The RNA molecule is less complex than the DNA molecule, since it usually is a single rather than a double chain. One can imagine that RNA might have been the means of transferring genetic information from a cell to its offspring prior to the evolution of DNA, and that the more capable DNA molecule eventually appeared and took over this function. There is a problem with this approach, however. Although it is conceivable that RNA was an earlier version of the molecule containing the primary genetic information, the RNA molecule cannot reproduce itself without the catalytic activity of enzymes. The enzymes are composed of proteins that are produced by using the genetic information contained within the RNA molecule, and the RNA molecule cannot be created without the catalytic activity of the enzymes. We are faced with a 'chicken or egg' dilemma.

This contradiction was somewhat resolved by the discovery of autocatalytic RNA molecules – RNA molecules that can, by themselves, act as enzymes to facilitate the production of new RNA molecules. Such 'ribozymes' (a word derived from the combination of ribonucleic acid – RNA – and enzymes) could have existed as an evolutionary remnant of earlier life forms that were simpler than modern-day life. However, even the simplest RNA molecule is very complex; it is unlikely that it would be created by the random combinations of organic molecules in a prebiotic soup. The likelihood of such a random occurrence (of either DNA or RNA by the random joining of molecules) has been likened to the implausibility of a tornado passing through a warehouse of airplane parts and leaving behind a fully assembled jumbo jet. However, just as evolution may have allowed DNA to supplant RNA as the molecule containing genetic information, and as modern-day RNA may have supplanted an earlier autocatalytic RNA, it seems plausible that the precursors of RNA might have been even simpler molecules that contained genetic information and were capable of reproduction.

So, we return again to the question of how the simplest possible molecules with the potential for containing some genetic information could be created and how they could reproduce. Regardless of the source of organic molecules, we need to think of ways in which the organics can be brought together to form a simple molecule that might be capable of reproduction and of transmission of the genetic information to their offspring. Such a molecule may be as simple as a polymer constructed by linking together simple molecules; recall that this is, in fact, what RNA and DNA are.

Such a polymer might have been created by any of several processes. For example, proteins (which include enzymes) might have been made by combining together individual amino acids. Alternatively, a simple precursor to an RNA molecule might be made by stringing together several nucleotides, which are the basic building block of both RNA and DNA. Finally, a polymer might be made by using some initial preexisting molecule or structure as a template for its production.

In the first possibility, the stringing together of amino acids is achieved by what is called a dehydration reaction. Two individual amino acid molecules can combine by forming a bond at a location in which one of them has an H atom and the other an OH complex. The result is the production of a polymer and the release of a water molecule. Because water is being driven from the polymer, the polymer is being dehydrated. This type of reaction can occur in environments in which there is a minimal amount of water, as the presence of water will tend to drive the reaction in the other direction – the water can combine with the polymer and yield two separate amino acids. If we start with a dilute organic soup, dehydration reactions might occur in shallow-water tidepools where the water will evaporate and leave behind the organic molecules. It can also occur in warm, relatively dry environments, again where the water has an opportunity to evaporate and leave the polymer behind.

In the second case, a primitive RNA molecule might have been constructed directly from its constituent molecules. The basic building block of the RNA and DNA molecule consists of a backbone composed of alter-

104   nating sugar and phosphate assemblages, with other molecules being attached to this backbone. The attached bases are the purines (adenine and guanine, or A and G in our shorthand) and the pyrimidines (cytosine, thymine and uracil, or C, T, and U). When heated, a mixture of ribose and purines will react together to form nucleosides, closely related to the nucleotide building block of RNA. A mixture of ribose and pyrimidines will not do so, however, making the production of the right building blocks impossible. Under the right conditions, certain molecules can catalyze the creation of both purine and pyrimidine nucleosides, but those conditions are very specific and probably did not apply for the prebiotic Earth. Might the earliest life have used only the purines to pass on genetic information, with the pyrimidines being added in later?

    Finally, is there something nonbiological capable of acting as a template for the creation of an RNA-like molecule? When RNA is created from DNA, or when a DNA molecule is copied, half of the DNA molecule serves as a template, onto which simpler molecules attach by chemical reactions, creating the new molecule. This likely pushes us back to the 'chicken and egg' problem, of needing to create a template first in order to create the molecule. Of course, the initial template did not have to be an RNA-like molecule. It might have been inorganic, such as the complex, semi-repetitive structure of various minerals. The mineral pyrite (FeS_2) has been suggested as a possible initial template, as has the clay mineral montmorillonite.

    Pyrite might make a viable template material because the mineral surface, at the atomic level, contains positive charges to which organic molecules can bond. A series of organic molecules attached to the pyrite surface might bond together to create more-complex, possibly RNA-like, molecules. In addition, energy is released in the creation of the pyrite molecule, in the form of free electrons (Figure 6.4), and these electrons can be used as the source of energy to trigger the combination of simpler organics into more-complicated ones. This might be a case of metabolism preceding reproduction as the triggering step in the formation of life.

    Alternatively, the surface of montmorillonite clay might be a place

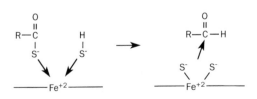

[Figure 6.4]
Schematic diagram showing how organic acids formed in solution might be reduced by the formation of pyrite on the surface of a mineral containing iron. This process could provide a source of energy capable of reducing carbon to create complicated organic molecules. (After Wachtershauser, 1988.)

where RNA-like molecules were first created. Clay molecules typically consist of layers of molecular structures, with the surface containing a repeating pattern of atoms. Organic molecules can bind to these sites, and possibly to each other. Essentially, the clay surface acts as a catalyst triggering the formation of repeating patterns of attached molecules. Recent experiments have demonstrated that RNA-like molecules can be made this way in solution.

This discussion has been couched more than any of the previous discussions in 'might have' and 'could have'. In reality, we do not know what process actually was the one by which the original biological entities were created. We imagine that RNA might have preceded DNA as the primary mechanism for passing on genetic information, and that some simpler, shorter molecule may have preceded RNA as a mechanism for passing on genetic information or for controlling the construction of new molecules. The mechanism of transition between these carriers of genetic information probably was very similar to Darwin-like evolution – survival of the fittest. But, the molecules that existed prior to RNA, and the chemistry responsible for their creation and for their evolution into more-RNA-like molecules, are unknown. Similarly, we imagine that DNA took over some of the functions of the earlier RNA – it is more efficient at being copied, at transmitting information, and in controlling the reproduction of a more complicated organism.

'Handedness' and membranes

There are two other issues that pertain to the origin of life that are not understood well either. They relate to the 'handedness' of the biologically relevant molecules and to the origin of membranes that serve to confine the contents of cells.

Molecules, as with many three-dimensional structures, can be either right- or left-handed. A right-handed molecule and a left-handed molecule are mirror images of each other, in the same way that a human right hand and left hand are mirror images (Figure 6.5). Because the forces controlling the chemical bonds made by atoms in a molecule are almost perfectly symmetrical, one would expect that right-handed and left-handed molecules should be present in equal amounts. In living organisms, however, both right- and left-handed molecules generally do not coexist. For example, the backbone of the RNA and DNA molecules contains sugars that exist only in their right-handed form, and amino acids used by life are

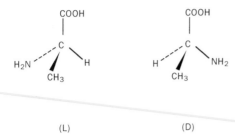

(L)   (D)

[Figure 6.5]
Right- ('L') and left-handed ('D') examples of
the amino acid alanine. The 'wedge' indicates
that the CH_3 lies above the plane of the figure,
while the 'dash' line indicates atoms below
the plane. These molecules are mirror images
of each other. Biological organisms utilize
only the 'L' version of amino acids and the
'D' version of sugars.

present only in their left-handed form. And, molecules of the wrong handedness cannot function in biological systems.

This handedness of life is confusing because all of the non-biological mechanisms that we discussed for making organic molecules as precursors to life will produce molecules with equal numbers of right- and left-handed versions. Additionally, organics that are brought to the Earth with incoming meteorites and comets usually contain nearly equal numbers of right- and left-handed molecules. Where did the handedness in terrestrial life come from?

Importantly, the formation of new RNA and DNA molecules from existing molecules cannot occur in the presence of equal numbers of right- and left-handed molecules, because the 'wrong'-handed molecules will keep them from forming properly. They can form only when their molecular components have the proper handedness. However, there is no reason to think that life could not exist using molecules of the opposite handedness, as long as all of the molecules were of the same orientation. Could the earliest life have contained some organisms that were left-handed and some that were right-handed, with competition, including random factors, ultimately favoring one over the other? Alternatively, could some competition between right- and left-handed molecules have favored one over the other prior to the origin of life, so that when life did form there would be only one handedness of molecule to incorporate?

The second issue deals with the presence of a membrane that encloses cells. The purpose of such a membrane is very straightforward – it isolates the contents of a cell from its surroundings, and it allows the charge separation necessary for redox reactions to occur (see Chapter 5). This allows a cell to keep its DNA and RNA together, distinct from that of other cells, to concentrate nutrients and chemical sources of energy that are required for survival, and to discard waste products by pushing them to the outside of the cell. The cell membrane would have to be unable to dissolve in water, but would need to be permeable to water and to other important molecules.

Cell membranes in modern life consist primarily of a double layer of

what are called phospholipids, with proteins intermixed within the layers.
Phospholipids are complex molecules similar to fats. They contain a hydrocarbon chain at one end that will not mix readily with water. At the other
end is a phosphate group that is partially charged and will be attracted to
water. The presence of a double layer allows these individual molecules to
line up with their hydrocarbon chains on the inside of the double layer,
away from any water, and with the phosphate groups on the outside. The
molecules are not strongly bonded to each other, so that nutrients, for
example, can pass through from one side to the other. The presence of proteins within the double-layer membrane aids in the transport across the
membrane; by using ATP as an energy source they can actively pump molecules across the membrane, or they can act as external bonding sites for
chemicals that can then trigger chemical reactions internal to the cell.

Although it is not known how cellular membranes originally formed,
two suggestions have been intriguing. First, small, spherical membrane
enclosures can form spontaneously by heating amino acids in water. The
amino acids form bonds between themselves and organize into a spherical
structure. These 'proteinoid microspheres' are typically 1 micron across
and are able to enclose organic molecules within their interior. Second, a
mixture of various organic polymers can, when heated, also form orga-
nized clusters that enclose a volume. These 'coacervates' can extend up to
500 microns across and also can isolate organic molecules from their surroundings.

These types of structures might have arisen spontaneously at about the
same time as the first reproducing molecules. They might have provided a
suitable means for concentrating organic molecules and instigating chemical reactions, and eventually might have been incorporated into the
reproduction activities of the molecules as they became cells. However,
modern cell membranes are not composed of materials similar to either of
these, and it is difficult to imagine how a transition to the modern membranes could have occurred.

The Earth's early history and the origin of life

Having come this far, is it possible to integrate the discussion of the possible mechanisms for the origin of life with some of the environmental constraints discussed earlier? Of particular interest are the environmental
effects of giant impacts (Chapter 2) and the evidence suggesting that the
oldest common ancestor of life was hyperthermophilic, thriving in water
that is much warmer than that at the surface of the Earth today (Chapter 5).

108   Recall that one of the most damaging effects of giant impacts during the earliest history of the Earth was the ability of the heat from the impact to evaporate some or all of the water in the oceans. Smaller impacts, say of objects only 100 km across, would evaporate the uppermost portions of the oceans. The largest impactors, up to 500 km across, would provide sufficient energy to evaporate essentially all of the oceans. Because smaller objects are usually more common than larger objects, the safest place for a primitive organism to reside would be at the bottom of the ocean. There, they would be the most protected from the effects of any but the largest, rarest impacts. The most risky place to subsist would be in shallow water, such as tidepools or small ponds. Clearly, this suggests that if life were trying to form in all of the possible environments that we have discussed, life originating in the deep ocean, as part of an ocean-floor hydrothermal system, would have the best chance of surviving.

The information discussed in the previous Chapter regarding the genetic distances between modern species suggested that the oldest common ancestor of terrestrial life was hyperthermophilic and probably existed in water at temperatures greater than 60 °C. Recall that this does not mean that life had to begin in such hot springs. The simpler scenario would have life originating in a hot environment and then beginning to migrate away from it. Of course, it is possible that life originated in a shallow-pond environment and then evolved to fit into a wide variety of environments, including ocean-floor hot springs. In the event of a giant impact, then, the biota that resided in the hot springs would have the best chances of surviving.

The fact that the global temperatures at the surface of the early Earth might have been as high as 85 °C probably does not change any of this discussion. As discussed in Chapter 4, such high temperatures may have resulted from the presence of a thick CO_2 greenhouse atmosphere. Most of the discussion of the formation of organic molecules and the possible mechanisms for the formation of life do not depend on temperatures being as clement as they are today, but can operate with higher temperatures as well.

Concluding comments

Where does this leave us? If understanding the origin of life on Earth requires that we reproduce the appropriate chemical environment and create life or molecules capable of life-like behavior in a laboratory, then we do not yet understand the origin of life. Further, many of the chemical reactions which have come the closest to generating molecules that may have been precursors of modern (or ancient) life require very specific conditions; these can be met in the laboratory but probably could not have been reached in the dirty environment of the prebiotic Earth.

This does not necessarily mean that the creation of life was an incredibly unlikely event, nor does it mean that we must resort to some extraterrestrial origin of life (be it panspermia or the intervention of a supernatural being). What it does mean is that we do not yet understand clearly the steps by which the first life originated. We do, however, understand the likely and plausible boundary conditions that were present at the origin of life. And, the advances during the past two or three decades in our understanding of the plausible chemical processes suggest that the origin of life was likely to have taken place in a very straightforward manner.

It is unlikely that we will ever fully understand the origin of life on Earth. The transition from nonliving, prebiological chemistry to the existence of entities fully capable of reproducing, of passing on genetic information, of changing through a 'survival-of-the-fittest' Darwinian evolution, and of metabolizing energy and using it to perform these functions, almost certainly was not a sharp transition. Rather, it must have involved a long series of steps, with only a few having been defined or described so far and even fewer have been demonstrated experimentally.

One concept that is very clear, though, is that life originated very quickly and evolved rapidly. Possibly as late as 3.9 billion years ago, conditions on Earth were not conducive to the continued existence of life. By 3.5 b.y.a., the fossil record shows the existence of organisms that appear identical to some that are still living today; these organisms would have utilized the molecule ATP to store energy, and DNA and RNA to pass on their genetic information during reproduction. For life to appear so quickly, and to evolve to a very sophisticated state so quickly, it must be very robust. It must be capable of forming easily in the right chemical environment. Ultimately, it is this apparent robustness that drives us to believe that life might have formed and might exist today on other planets in our solar system or on planets around other stars.

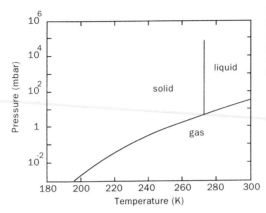

[Figure 7.2]
Phase diagram for water. Stability fields for the solid, liquid, and gas phases are indicated. At typical martian temperatures, liquid water is not stable.

results in a lower density, and the ice floats. Because of this, ice can be present floating on the oceans in the polar regions and can be stable over a wide range of surface temperatures. If the planet were to cool off, a little more ice would form in response; and if it were to warm up, some of the ice would melt. If ice were to sink instead of floating, a planet that is cooling off would begin to form ice, and that ice would sink. This would expose more water at the surface, allow the freezing of more water into ice and its sinking, and the entire global oceans would quickly freeze.

Water is a unique substance in this regard, allowing all three phases – solid, liquid, and vapor – to coexist under a wide variety of climatic conditions (Figure 7.2). Ammonia, methane, and ethane do not have this property – given that the ices will sink in the liquid, it would be hard to imagine a scenario where the three phases could coexist. And, while oceans or lakes of these liquids might exist (as, for example, on the Saturn satellite Titan), they would not be as stable over a wide range of temperatures.

There is another property of water that plays a very important role in terrestrial life, but may not be as important elsewhere. Water is a polar molecule, meaning that its intrinsic charges (the negatively charged electron and the positively charged proton) are not symmetrically distributed. Other polar molecules can dissolve in water, while non-polar molecules (those with symmetrically distributed charges) cannot. For example, oil, or any long-chain biological hydrocarbons, will not dissolve in water. In terrestrial life, this fact is extremely important – if the membranes or walls that surround cells were polar, water would be able to break them apart. Because these membranes serve to separate the inside from the outside of cells, they must remain intact in order for life to function. On Earth, they are made of long-chain non-polar hydrocarbons, ensuring that they will be stable in water.

On another planet, life could not evolve in the presence of liquid water without having a membrane or cell wall that consisted of non-polar molecules. On the other hand, however, suppose another planet had oceans of the non-polar molecules methane or ethane. Earth-like membranes of non-polar molecules would dissolve in them, inhibiting the ability of life to function. However, the non-polar hydrocarbon chains can be made into polar molecules simply by the addition of H and OH complexes at the right locations. There is nothing about such polar chains of hydrocarbons that would rule out the other functions of living organisms. Therefore, the presence of non-polar liquids would not rule out life, but any life that evolved in their presence would have a fundamentally different chemical structure than terrestrial life.

Despite this ability for life to exist in liquids such as methane or ethane, the other factors argue that water is a more-suitable liquid for supporting life. Water on another planet could be present in any of a variety of locations. It could be present as an ocean (as on the Earth), as isolated lakes and ponds, or as groundwater. If it is present at the surface, then there are requirements placed on the environment – the planet must be warm enough for the water to melt, yet not so warm that it all would evaporate into the atmosphere. There is an alternative possibility – the water could be present below the surface, in hydrothermal systems heated by volcanic activity. This option places few requirements on the climate or environment. Water is a common component of planets as they form, and is released readily through volcanic activity. This volcanic water, even if it does not rise all the way to the surface, is available within the subsurface and can be heated by internal volcanic activity; thus, it is available for use by biota. Of course, subsurface water and a subsurface biosphere would not be readily observable by spacecraft or through telescopes.

In addition to water, life requires the presence of all of the elements that participate in relevant biochemical reactions. Remarkably, of the 92 naturally occurring elements in the universe, only 21 play a major role in terrestrial life. The major biogenic elements are carbon (C), hydrogen (H), oxygen (O), nitrogen (N), sulfur (S), and phosphorous (P). Other important elements include sodium (Na), potassium (K), magnesium (Mg), calcium (Ca), and chlorine (Cl). Biogenic trace elements are manganese (Mn), iron (Fe), cobalt (Co), copper (Cu), and zinc (Zn), with some organisms also utilizing boron (B), aluminum (Al), vanadium (V), molybdenum (Mo), and iodine (I). Other elements do participate in biology, but at a lower level of significance. For example, silicon (Si) is abundant in many unicellular organisms such as diatoms and radiolarians, nickel (Ni) plays a catalyzing role in various organisms, and bromine (Br) is used by a number of marine organisms.

For comparison, the most abundant elements in the universe are, in order of decreasing abundance, H, He, O, C, N, Ne, Si, Mg, Fe, S, Ar, Al, Ca, Ni, and Na. Similarly, the most abundant elements in the Earth's crust are O, Si, Al, Fe, Ca, Na, and K. Clearly, life has not chosen to take advantage of those elements that are most abundant but, rather, has utilized elements that must have chemical properties that are advantageous. It turns out that the four most significant biogenic elements, H, O, N, and C, are able to achieve electronic stability by chemical reactions that allow them to gain or lose 1, 2, 3, and 4 electrons, respectively. Further, each one is the smallest atom that can do this. The small size allows these atoms to form the strongest chemical bonds; other elements might be able to form some of the same chemical bonds, but their larger size means that they would be less stable. Also, C, O, and N have the ability to form double and triple bonds, allowing them to form a wider variety of molecules.

Silicon is often suggested as a plausible alternative to carbon in forming a central basis for life, due to its ability to form long carbon-like chains. Although Si is not as abundant as C in the universe, it is much more abundant in the Earth's crust. Can it serve this purpose? Probably not. Silicon cannot form double bonds the way that carbon can. As a result of carbon's ability to form double bonds, it can combine with two oxygen atoms to form CO_2. In this form, the carbon can exist as a gas and can dissolve in water. Without this ability to move through the atmosphere and the oceans, carbon would not be readily available to organisms. Si–O bonds, on the other hand, are single bonds. As a result, silicon and oxygen can combine together, but they form extended structures such as the three-dimensional mineral quartz. Silicon does not combine with any other elements to form a compound that can exist as a gas at the surface of the Earth. In fact, there are no conditions where it could exist as a gas in substantial quantities that could also allow life to exist. And, although silicon can dissolve in water, it does so in such small amounts that it would not be available in any useful quantities.

There is another problem with silicon. Although Si will form chains in a manner similar to C, the bonds between Si atoms are looser and weaker than are the bonds between C atoms. And, because of the availability of sites for bonding with another atom, a chain molecule based on the silicon atom is much more susceptible to chemical interactions with oxygen, ammonia, or water, which would break it apart. For these reasons, Si is much less suitable as a basis for life than is C.

Carbon is found not only on the Earth, but also in meteorites, in interstellar space, and in other stars. It is so widespread in the universe that it would be a ready basis for life elsewhere. **115**

Energy must be available in a form that can be utilized by organisms to drive biochemical reactions. If we look at the possible locations for the origin of life on Earth or where life on Earth can exist today, we see a variety of forms of energy that could be or are tapped. These include sunlight, through photosynthesis; lightning or electrical discharge, through the metabolism of organic molecules created by the lightning; heat in hydrothermal systems, through the use of organics or other reduced species created by chemical interactions; or chemical energy, by oxidizing elements in the environment, including many of the elements in the crust or hydrogen gas or methane obtained by interactions with the crust.

What sorts of planetary environments are likely to provide access to energy in these usable forms? A remarkably wide variety of possible environments would work. Geothermal energy should be widely available on solid planets. The ultimate source of the energy is the heat from the formation of the planet and from the decay of radioactive elements within the planet's interior. (On Io, a satellite of Jupiter, there is an additional heat source from the dissipation of tidal energy; this will be discussed in Chapter 13.) This energy is available on Earth primarily through the movement of water within the crust, circulating between regions of differing temperatures. The disequilibrium that results from the temperature changes drives chemical reactions.

We would expect that any planet undergoing active geological processes would have access to geothermal energy. This includes any planet that is big enough to retain heat for long periods of time, rather than cooling off quickly by conduction to space. How big is big enough? Even the Earth's Moon is big enough to have had volcanic activity at the surface until more than a billion years or so had passed since its formation (of course, the Moon is too small to have retained any significant atmosphere or hydrosphere). The Moon, however, is small enough that it now has lost the bulk of its heat, and it has not been volcanically active near its surface for, probably, around three billion years (Figure 7.3). It takes a planet as big as Mars to remain active for 4.5 billion years and to retain a substantial atmosphere. Clearly, a planet as big as Mars or bigger is intrinsically capable of supporting life, depending on its climate history.

The presence of an atmosphere and hydrosphere is important, as well, for processes involving lightning or ultraviolet light. Both of these energy sources can act as drivers for chemical reactions in planetary atmospheres.

Chemical energy would have been available to early terrestrial organ-

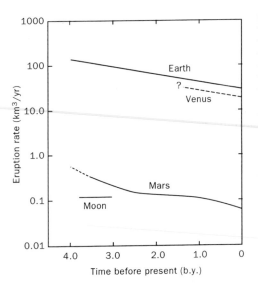

[Figure 7.3]
Volcanic eruption rates on the terrestrial planets through time. All rates are for intrusive and extrusive eruptions together, assuming that the total is 8.5 times the extrusive rate alone. The Earth eruption rates during the past are calculated from the best estimate of the current eruption rate by assuming that the volcanic eruption rate is proportional to the rate of generation of heat within the interior by radioactive decay. The present-day Venus eruption rate is an upper limit, and the actual value might be several times lower; it is extrapolated into the past the same way as for the Earth, with the dashed line indicating the uncertainty associated with the possibility that volcanic resurfacing on Venus might occur intermittently and catastrophically, with quiescent periods interspersed. The Mars history is based on observed rates as a function of time, using the best estimate for total volumes and for the ages of geologic units; rates for surfaces older than 3.5 b.y.a. are uncertain due to the difficulty of recognizing volcanic surfaces. The lunar rate is an average, and was determined from the total volume of volcanic material, assuming that eruptions spanned the time from 3.8 to 3.1 b.y.a.; there is a negligible volume of volcanic material younger than this.

isms simply by metabolizing the organic molecules that would have been present in the environment. Organic molecules should be accessible as a result of formation within the atmosphere or at hydrothermal systems or by supply from interplanetary space (see Chapter 6). Metabolization of other reduced molecules, such as H_2S or H_2, also is possible. These molecules would be associated with volcanic centers or hydrothermal systems and could be produced even by chemical interactions between water and minerals in the ambient crustal environment.

Finally, there is an issue that has been alluded to a number of times but which we have not addressed explicitly – the stability of the environment. This refers to the geologic stability, the climatic stability, and the absence of disrupting influences. We saw the significance of this in the discussion of the effects that giant impacts can have (Chapter 2). They can cause sudden and dramatic changes in the temperature, availability of water, and availability of sunlight. Clearly, life could not survive continuously on Earth until the impact rate had declined substantially from its early values. Even then, the occasional large impacts that have occurred throughout geologic time have had substantial impact on the environment. In our solar system, the impact rate has been moderated due to the gravitational influence of the outer planets (see Chapter 2); in another solar system that did not have a Jupiter or Saturn, an Earth-like planet might be too heavily bombarded to sustain life.

An additional disruptive influence for life would be that due to ultraviolet light from the Sun. Light at wavelengths shorter than about 0.3

microns is able to break the molecular bonds in organic molecules (visible light has wavelengths between about 0.4–0.7 microns). In fact, exposure to the full solar ultraviolet flux can quickly kill any organisms and sterilize any surface. This has been demonstrated, for example, by the NASA Long Duration Exposure Facility, an Earth-orbiting spacecraft designed to determine the effects of exposure to space; after sitting in space for a long period, the outer surfaces were quite effectively sterilized. Closer to home, we all are familiar with the damaging effects of exposure to the Sun's ultraviolet light – it can disrupt the organic molecules in our skin and trigger mutations that lead to cancer.

Although the total amount of energy coming from the Sun was 30% lower early in its history, its output of ultraviolet light at that time was several orders of magnitude greater. In the present Earth atmosphere, gaseous ozone (O_3) in the stratosphere absorbs most of the damaging photons and protects the surface; only small amounts of ultraviolet light get through. On the early Earth, organisms would have been partially protected by the presence of a very thick atmosphere, even in the absence of ozone; however, organisms that live beneath only a very shallow layer of either dirt or water would be quite effectively shielded. (It is interesting that terrestrial life did not migrate to land surfaces until after the amount of oxygen in the atmosphere had increased; the presence of abundant oxygen would have led to the creation of an ozone layer.) Any planet that had liquid water at the surface, or in which the organisms lived beneath the surface, would be protected from ultraviolet energy. Of course, organisms that live entirely beneath the surface are, again, very difficult to detect.

Concluding comments

The issues discussed above would seem to suggest that the conditions that we find on the Earth are the most suitable and appropriate for life. This would include, especially, the necessity of having liquid water and the use of carbon as a basis for life rather than, say, silicon. While there are legitimate reasons to think that these are necessary conditions rather than merely sufficient conditions, we have to be careful not to read too much into them. Because we have only the one example of life in the universe, on Earth, it is extremely dangerous to make generalizations about the conditions needed for life. And, there are instances in terrestrial life where the simplest solution to various problems has not always been chosen. We should keep an open mind to the possibility that life elsewhere might be different from that on Earth. Importantly, keeping an open mind will not limit the possibility for life elsewhere, but will enhance it. For example, although water is the most abundant volatile in the universe, there are places where other liquids might exist instead (such as the surface of the Saturn satellite Titan). Thus, allowing for other liquids, or other chemistries, merely adds to the number of locations where life might conceivably exist. However, we will take the conservative approach, for the most part, and look for places where conditions that we know gave rise to life might exist.

We now can begin to apply our conclusions about the conditions required for life to a discussion of where life is likely to form and where it might continue to exist up to the present. We will focus first on our own solar system, looking for places where liquid water might exist today or might have existed in the past, and where conditions might be conducive to the occurrence of life. Of course, the most obvious place to look will be Mars, where liquid water has existed at least occasionally on the surface throughout geologic time, but there are other planets in our solar system that are of interest as well. Outside of our own solar system, we will look for planets that exist around other stars, and discuss how abundant they are, what their properties might be, and whether they might be capable of sustaining life.

At the present, we know of no life on Mars. Despite this, there is the possibility that life might have arisen at some time in the past and might either have gone extinct or still exist somewhere on the planet today. In this chapter, we will discuss the reasons why living organisms could have originated on Mars independent of the Earth or been transferred there from the Earth, and why they might have survived up to the present. The recent finding of possible fossil remnants of martian life in one of the meteorites that came from Mars will be discussed in Chapter 9.

The present-day surface of Mars

Global imaging of the martian surface from spacecraft provides evidence that a wide variety of geological processes have operated over time. After Earth, Mars may have the most variety in the solar system in the types of processes that have acted. There is abundant evidence for volcanism, tectonism, wind-related movement of material, ice- and water-related geology, and large-scale impacts, for example. At a smaller physical scale, pictures taken from landed spacecraft show a desert-like landscape, with small drifts or dunes of sand-sized or smaller particles, and annual dust storms that can deposit or erode thin layers of dust at the surface (Figure 8.1).

The surface of Mars today is generally not seen as being conducive to the presence of widespread life. Although the peak daytime surface temperatures near the equator can rise above the freezing point of water during much or all of the year, the average surface temperature is only about 220 K (-53 °C), substantially below the freezing point. The atmosphere is relatively thin, averaging about 6 mbar and consisting primarily of carbon dioxide. By way of comparison, the Earth's atmosphere has a pressure of 1013 mbar (1.013 bar) – more than 100 times the thickness of the martian atmosphere – and is composed primarily of nitrogen and oxygen. The amount of CO_2 in the Earth's atmosphere, about 340 parts per million, is equivalent to a pressure of about 0.3 mbar.

Mars has days and seasons similar to those on the Earth. The martian day lasts 24.6 Earth hours, which is just over a terrestrial day. The distance from Mars to the Sun (1.5 A.U.) is greater than the distance from Earth to the Sun, so that a Mars year is longer than an Earth year; the martian year is 686 Earth days or 669 Mars days. The tilt of Mars' polar axis relative to its orbital plane is 25°, compared with the Earth's 23.5°, so the nature of the seasons is somewhat similar. The martian orbit is not as circular as the

[Figure 8.1]
Image of the martian surface, taken from the *Viking-1* lander in 1976. The large rock on the left is about 2 m across. There is abundant evidence for windblown removal and deposition of material. (NASA photo.)

Earth's orbit, however, so its seasons are more extreme: Mars is closest to the Sun during summer in the southern hemisphere and farthest from the Sun during summer in the northern hemisphere. As a result, the southern seasons are more extreme (warmer in summer, colder in winter) and the northern seasons are more moderate.

The amount of carbon dioxide gas in the martian atmosphere varies throughout the seasons. During the winter season, the surface and atmosphere at high latitudes become cold enough for the CO_2 in the atmosphere to condense out as frost, forming a seasonal polar ice deposit; during the summer, the frost sublimes back into the atmosphere. (The martian atmospheric pressure is low enough that liquid CO_2 cannot exist, so frost will sublime directly from solid to vapor.) As much as 1/3 of the atmospheric CO_2 cycles seasonally between the north and south polar caps, as frost deposits. The seasonal polar CO_2 frost deposit can be up to about a meter thick, and will extend to latitudes as low as 40°. The seasonal

[Figure 8.2]
Polar projection of *Hubble Space Telescope* images of Mars, showing the retreating north-polar seasonal CO_2 polar cap over a period of six months. The left image is early northern-hemisphere spring, the middle image is mid-spring, and the right image is early-summer (and shows the residual polar cap after the seasonal frost has disappeared). (NASA photo courtesy of P.B. James, R.T. Clancy, and S.W. Lee.)

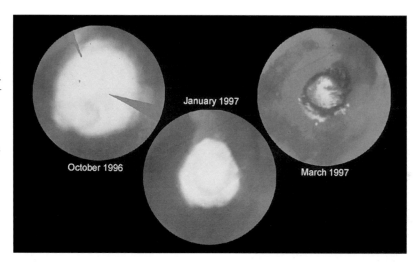

January 1997

October 1996

March 1997

advance and retreat of the polar caps that results from the deposition and removal of frost on the surface has been observed both from spacecraft in orbit around Mars and from telescopic observations (Figure 8.2).

During the northern-hemisphere summer season, spacecraft observations show that all of the CO_2 frost on the surface disappears. Beneath it poleward of about 80° latitude is a relatively bright, white polar deposit that does not disappear. This permanent polar cap is a mixture of water ice and some red martian dust (Figure 8.3). This residual water-ice cap is able to heat up during the summer season, to perhaps as warm as 205 K; this is sufficient to allow some water vapor to sublime into the atmosphere, but not warm enough to allow the ice to melt. Interestingly, the south-polar CO_2 frost does not seem to disappear completely during the summer, so a south-polar water-ice cap, if present, is never revealed. The differences between the two polar caps are only partly understood, and appear to result from differences in elevation, the influence of atmospheric dust on the heat balance of the frost, and effects of subsurface conduction of energy.

Water is abundant on Mars away from the polar caps as well. Water vapor and ice crystals are present in the atmosphere, supplied by the summertime sublimation of the north-polar water-ice cap. In fact, because of the cold temperature of the atmosphere, water is often saturated in the atmosphere or near the surface, resulting in the presence of clouds. Water ice almost certainly is present within the ground at high latitudes; there, the subsurface is cold enough within the top few meters that atmospheric water vapor will diffuse from the atmosphere into the surface and condense as ice.

Liquid water is not stable at the martian surface (see Figure 7.2).

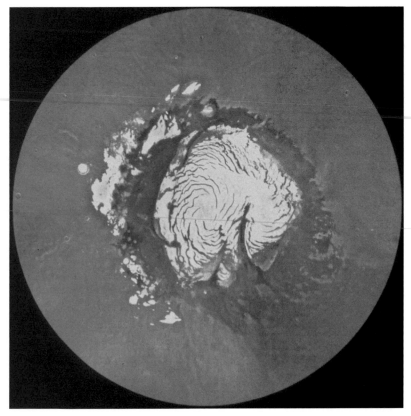

[Figure 8.3]
Viking composite image of the martian residual north-polar ice cap. The bright material is predominantly water ice, with some dust mixed in. (NASA photo.)

Although temperatures can occasionally rise above the melting point, liquid water from melting ice would quickly evaporate into the atmosphere. Subsurface temperatures are below freezing and would keep any water frozen as ice, though isolated pockets of liquid water still might exist. The presence of salts, of the right composition and in sufficient quantity, can lower the freezing point enough to allow ice to melt at lower temperatures; salt-rich water still would evaporate into the atmosphere, however, so that such a brine would not be stable. Ice crystals trapped within closed pores in rocks or regolith grains might be able to melt and still be inhibited from evaporating; if it occurs at all, however, such liquid likely would be rare.

In addition to the lack of liquid water at the martian surface, other aspects of the environment make it less than conducive to the existence of life. The atmosphere is thin enough and contains a small enough amount of ozone that ultraviolet light from the Sun can reach the surface almost unattenuated over most of the planet. There is some ozone in the winter hemisphere atmosphere; however, it can only absorb some of the ultraviolet light, during only part of the year, and only over a fraction of the planet.

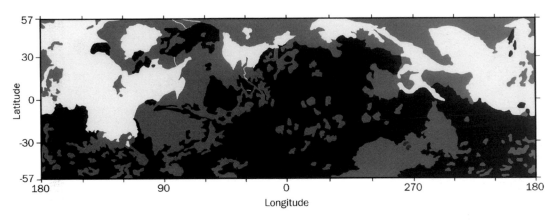

[Figure 8.4]
Simplified geologic map of Mars, showing the different ages of the surface. The dark gray regions are Noachian in age (corresponding roughly to ages from about 4.0–3.5 b.y.a.), the light gray regions are Hesperian (3.5–1.8 b.y.a.) and the white areas are Amazonian (younger than 1.8 b.y.). Roughly half of the surface is older than 3.5 b.y. in age. (After Tanaka, 1986.)

Based on the existing measurements at the martian surface, Mars appears to be a very hostile location for life. If life did exist, it would have a very difficult time obtaining water in any significant quantities and it could easily be harmed by ultraviolet light.

A warmer, wetter climate on ancient Mars

The climate on Mars may not always have been so detrimental to life, however. Early in its history, the climate is thought to have been more capable of allowing liquid water to exist at or near the surface. The evidence comes from interpretation of images that show the geology of the surface features. A substantial fraction of the surface of Mars is estimated to be older than around 3.5 billion years, based on the number of impact craters that are present (Figure 8.4). The presence of these ancient surfaces provides a valuable window into the planet's early history. In particular, two characteristics of these older surfaces suggest that the climate prior to about 3.5 b.y.a. was different somehow from the present-day climate.

First, impact craters smaller than about 15 km in diameter have been obliterated on these oldest surfaces, and impact craters larger than this have undergone substantial erosion (Figure 8.5). By comparison, impact craters that are found on surfaces that are younger than 3.5 b.y. have not been altered significantly. This suggests that erosion rates were greater by a factor of about 1000 early in martian history. The style of erosion that is seen on some of the remaining larger impact craters, with gullies on the walls of craters, for example, is indicative of surface runoff of water. Erosion by water also is thought to be responsible for having destroyed the smaller craters.

[Figure 8.6]
Dendritic valley networks on
the ancient martian surface.
These are thought to have
formed by the steady flow of
water or water-rich debris, and
the climate when they formed
must have been more clement
than today. (NASA photo.)

Second, many of these same older surfaces contain networks of valleys
that form dendritic patterns similar to terrestrial water-carved streams
(Figure 8.6). There is some continuing debate as to exactly how these
valleys were formed – the process may have involved runoff of precipitation, formation by seepage of subsurface water in a process termed
'sapping', erosion by water-rich debris flows, or a combination of these
processes. Independent of the exact process, their formation must have
involved the presence of liquid water at or very near to the surface during
these earlier epochs.

In combination, then, the geologic evidence suggests that the martian
climate prior to about 3.5 b.y.a. was somehow warmer than the present
climate and allowed liquid water to exist at the surface more readily or in
greater abundance than it does today. Unfortunately, the observations do
not allow a unique interpretation of what the temperature, atmospheric
pressure, or partitioning of water between the surface, subsurface, and
atmosphere were at that time. Also, the mechanism for producing a
warmer climate is uncertain. As with the early Earth, a thick CO_2 green-

[Figure 8.5]
The ancient, heavily cratered
highlands on Mars. Notice that
the large impact craters have
been substantially eroded and
that there is a lack of smaller
craters. The image is about
900 km from top to bottom.
(NASA photo.)

house might have warmed the atmosphere. However, because Mars is farther from the Sun than is the Earth, the amount of CO_2 required to raise its temperature enough to allow liquid water may have saturated the martian atmosphere; saturation and condensation would have kept the CO_2 in a greenhouse from being abundant enough to provide much heating, although small amounts of atmospheric dust or other gases may have alleviated this problem.

A transition from a warmer, early climate to the colder present climate may have resulted from loss of atmospheric gases to space. Evidence from measurements of martian stable isotopes suggests that a large fraction of the volatiles from early Mars might have been lost to space. In addition, the large impacts that occurred, and which are recorded in the ancient surfaces as the remaining impact craters, would have ejected gas from the atmosphere to space. Together, these two loss processes may account for

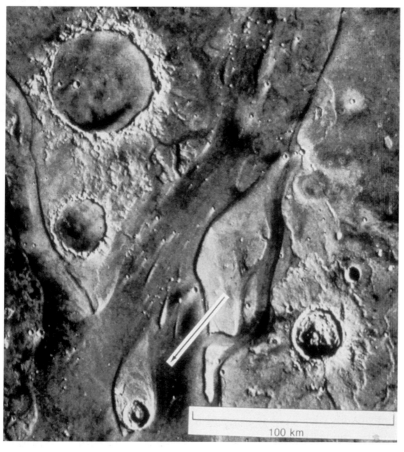

[Figure 8.7]
Well developed flood channels, carved by water flowing over the surface. The channel is over 50 km across in places; it can be seen to diverge around obstacles, leaving a 'streamlined' island in its wake (arrow). (NASA photo courtesy of the Lunar and Planetary Institute.)

100 km

[Figure 8.8]
Source region for some of the catastrophic floods. The channel originates from a region of chaotic debris, with the source for the water being beneath the surface. The scarp at the head is about 20 km across and 2 km deep. (NASA photo.)

much of the gas that would have been present in the early greenhouse atmosphere.

Despite the change in climate, water has been geologically active on Mars even since these earliest times. There are flood channels that appear to have been carved by water released in large, catastrophic events (Figure 8.7). These channels show morphologic features similar in shape to those seen in terrestrial catastrophic floods. These include shorelines carved at the edge of the channels, aerodynamically shaped erosional remnants where the water flowed around obstacles resistant to erosion, features analogous to sandbars that are kilometers in size, and chaotic collapse deposits located near the source region of the water. These flood events do not flow from large lakes or drained reservoirs; rather, they appear to flow out from beneath the surface, indicating release of water that was trapped in the martian subsurface (Figure 8.8). Even though the surface tempera-

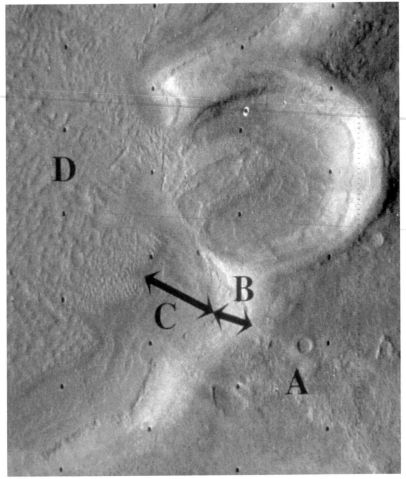

[Figure 8.9]
Possible shoreline features intruding into an impact crater. The series of benches that appear to run parallel to topographic lines may have been carved by waves that occurred on the surface of a large standing body of water. (NASA photo courtesy of T. Parker.)

ture is too low to allow water to melt, temperatures at kilometer depths will be higher because of the presence of geothermal heating deep within the planet; the warmer temperatures will allow ice to melt and liquid water to exist only a kilometer or two beneath the surface. The flood events appear to have occurred throughout martian history, indicating that water was available in the martian crust for much or all of its history. Most of the floods drained into the northern lowlands, where the water may have ponded up as standing bodies of water before it froze, percolated into the subsurface or evaporated into the atmosphere. There is even the possibility that the water might have been liquid for a long enough time that waves could have carved out shoreline features in the surrounding regions (Figure 8.9).

[Figure 8.10]
Olympus Mons shield volcano. The volcano is about 600 km across at its base, and stands about 25 km high at its peak. The caldera complex is almost 100 km across. (NASA photo courtesy of Lunar and Planetary Institute.)

Water also has been released throughout geologic time by volcanic activity. There is abundant evidence for volcanism on Mars. The northern plains are volcanic, covering about a third of the martian surface. In addition, there are numerous large volcanoes, standing up to 25 km above the surrounding plains (Figure 8.10). Because Mars has not undergone sufficient erosion or geological activity to destroy the older surfaces, we can see the different geologic processes that have acted throughout martian history. These indicate that volcanism has been active during most of martian history. Although the oldest surfaces do not show features suggesting that they are volcanic, some people expect that volcanism was very active then, because of a greater availability of interior heat, and that the

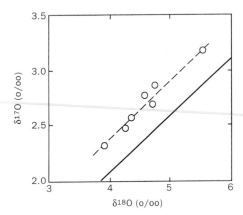

[Figure 8.11]
Comparison of the oxygen isotopes in minerals from the martian meteorites with those from minerals from Earth. The x axis shows the ratio of $^{18}O/^{16}O$ (as the deviation of that ratio from the ratio in a standard sample, in parts per thousand), and the y axis shows the ratio of $^{17}O/^{16}O$. All terrestrial (and lunar) rocks fall along the solid line, within the measurement error. The open circles represent measurements from the martian meteorites, and the dashed line is the best fit through them. Notice that the martian line clearly is offset from the terrestrial line; this indicates that the meteorites cannot be from the Earth. (After Karlsson et al., 1992.)

diagnostic features may have been eroded away. Similarly, we do not know if Mars is volcanically active today, but the geologic evidence suggests that it has been active at least up until the most recent epochs.

Evidence for very recent volcanism is seen in meteorites collected on the Earth and thought to be from Mars. These meteorites are known as the SNC (pronounced 'snick') meteorites, after the initials of the three type examples, Shergotty, Nakhla, and Chassigny. They are pieces of volcanic rock that crystallized between 1.3 b.y.a. and 180 m.y.a. They are thought to be of martian origin for several reasons: as young volcanic rocks, they must be from a planet that is large enough to have had active volcanism relatively late in history; this means one of the large terrestrial planets. The oxygen isotopes (the ratios of ^{18}O, ^{17}O, and ^{16}O) absolutely rule out an origin on the Earth or on the Moon (Figure 8.11). This leaves Mars and Venus as the most likely source planets. The most convincing argument for a martian origin is that two of the dozen or so martian meteorites contain a component of gas trapped within them that is identical in composition to the Mars atmosphere and distinct from any other source of gas in the solar system (Figure 8.12). The gas is thought to have been implanted into the rocks by the shock generated from the impact event that ejected the rocks from the martian surface into space. Some of these volcanic rocks crystallized from molten rock as recently as 180 m.y.a. Clearly, there has been at least some active volcanism in the very recent past. It is unlikely that this was the very last gasp of a dying planet, suggesting that Mars is likely to be volcanically active today.

Some of these rocks also show evidence for the presence of water. They contain some residual amounts of water that was not released into the atmosphere during the volcanic eruption that created them. Some of them

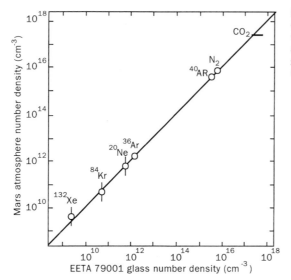

[Figure 8.12]
Comparison for several gases of
their abundance in the martian
atmosphere and trapped within the
glass in the martian meteorite
EETA79001. (After Pepin, 1994.)

show geochemical evidence for alteration by hot water flowing through the rocks subsequent to their crystallization – that is, evidence for hydrothermal systems. Certainly, given the obvious evidence for the presence of heat from volcanism and from impacts and for water within the crust and at the surface, we would expect hydrothermal systems and hot springs to exist. Some of the large volcanoes even have channels on their flanks that appear to have been carved by water (Figure 8.13). Again, this suggests the presence of hot springs in which water associated with the hydrothermal systems was actually released to the surface.

Clearly, water has been a significant species at and near the martian surface throughout its history. The early climate included the presence of water at the surface, and water has been present within the subsurface ever since. Although there is substantial debate as to the exact amount of water, there could have been enough to form an ocean a kilometer thick if all of the water were present at the surface at the same time.

Could life have formed on Mars?

As described earlier, life on Earth formed sometime prior to 3.5 b.y.a. and probably before 3.85 b.y.a. The details of its origin are not known, and even the mechanisms of origin are uncertain. Possible sites for the formation of

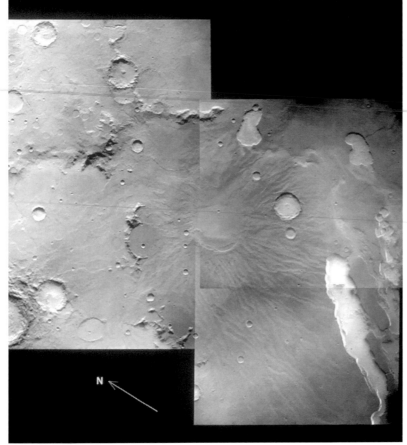

[Figure 8.13]
The volcano Hadriaca Patera, with the catastrophic outflow channel Dao Vallis on its flanks. The volcanic caldera can be seen at the center, and channels carved on its flanks move downhill, outward in all directions. Dao Vallis, at the bottom right, probably was carved by water released or mobilized during the eruptions. (NASA photo courtesy of M. Mellon.)

life include transient evaporating ponds, hydrothermal vents where water circulates beneath the surface near volcanic intrusions, or the surfaces of clay minerals that could provide stability and order to long chains of molecules. The requirements for the formation of life are a source of organic molecules, a source of energy that can drive disequilibrium processes, and access to the biogenic elements (such as C, H, O, N, S, and P). The source of organics could be external (for example, formed in interplanetary space and supplied to the Earth along with meteoritic dust and debris that accreted onto the Earth) or internal (formed by chemical reactions within the terrestrial atmosphere or at hydrothermal vents in the oceans).

Mars appears to have had all of these same necessary conditions for the formation of life.

The climate on early Mars may have been somewhat similar to the climate on Earth at the time of the origin of life on Earth. Although martian erosion rates were undoubtedly substantially less than terrestrial erosion rates (which suggests less-widespread water), liquid water certainly was present on both planets. Both planets probably had a mildly reducing atmosphere, containing substantial quantities of carbon dioxide. If life on Earth formed in a relatively quiescent, near-surface environment, then it is quite possible and even plausible that life could have formed on the martian surface at the same time. If such were the case, then an early, globally distributed surface biosphere on Mars may have existed.

Alternatively, terrestrial life could have formed in a hot-spring environment; such environments must have been widespread on early Mars as well. A martian hot spring or hydrothermal system would require substantial sources of heat and water in the crust. Heat would have been retained within the subsurface from the impacts of accreting planetesimals. Such impacts were a common occurrence during the tail end of the heavy bombardment, as recorded in the impact craters on the oldest surfaces. Volcanism on Mars also was capable of supplying heat. Both extrusive and intrusive volcanism have occurred throughout martian history, although the amount has been declining at least during the latter billion years. Clearly, there was sufficient heat to drive hydrothermal circulation or hot springs.

The geologic evidence also suggests that abundant water has been present within the crust of Mars. Given the widespread evidence for both heat sources and accessible water, it is likely that subsurface hydrothermal systems have occurred throughout martian history. If life could form in such an environment, then there is every reason to believe that life might have formed on Mars.

An alternative source for life on Mars might have been the Earth itself. Impacts of asteroids onto the surfaces of the terrestrial planets are capable of ejecting surface rocks into space. Once in space, gravitational encounters with their planet of origin would alter their orbits. The orbits of ejecta from the Earth could have altered over time so that they cross Mars' orbit; similarly, the orbits of ejecta from Mars could evolve to the point where they cross Earth's orbit. At that point, collisions can occur and there is a net transfer of mass from one planet to another. The martian meteorites, which are from Mars and were collected on the Earth, demonstrate that this interplanetary exchange process can occur. Not all of the rocks and debris ejected into space by an impact will be heated or shocked substantially, and bacteria or bacterial spores that might be contained within them would be able to survive the ejection event; if they survive the travel through space and if they land in a satisfactory environment on their new planet, they could survive and multiply. Of course, we can always ask the question, then, as to which planet served as the location for the origin of life – could life have originated on Mars and been transferred subsequently to the Earth? This question is of profound significance, but is not easily answered.

If life ever existed on Mars either by formation in an independent origin or by transferal from the Earth, is it possible that it could have survived up to the present? Although life may not be globally distributed on Mars today, as inferred from the *Viking* lander spacecraft experiments, it could continue to thrive sequestered in the occasional localized ecological niche. Such niches could be liquid water or hot springs associated with extrusive and intrusive volcanism or buried deeply beneath the surface where liquid water could still exist.

Although the rate of volcanism has been declining throughout the latter half of martian history, it probably has occurred all the way up to the present. The abundance of water within the crust suggests that recent surface volcanism would have hot springs or near-surface hydrothermal systems associated with it; life could thrive in such environments.

In addition, life could exist deep within the crust, where widespread liquid water could occur. The heat still present within Mars' interior ensures that ground ice would melt near the equator at depths as shallow as a couple of kilometers. The presence of water within the crust is suggested by the occurrence of the catastrophic flood channels, wherein substantial quantities of water appear to have been occasionally released to the surface from the deep crust throughout geologic time. The recent discovery of terrestrial organisms living deep within the Columbia River basalts in the Pacific Northwest (see Chapter 5) bolsters the possibility of organisms living in these buried environments on Mars. These organisms survive by metabolizing the hydrogen produced by chemical interactions between basalt and water in the pore spaces; they are completely independent of any input of chemical energy from the surface, and survive completely isolated from the surface. Of course, these terrestrial organisms did not originate in the subsurface but must have migrated there from the surface. Similar migration to the deep subsurface could have occurred on Mars as surface temperatures declined from early higher values to their present cold level.

The search for life

But didn't the *Viking* spacecraft in the late 1970s rule out the presence of life on Mars? We need to discuss the *Viking* lander biology experiments in some detail in order to be able to evaluate them in light of our present understanding of the possible occurrence of martian life.

The *Viking* spacecraft consisted of two orbiters and two landers, launched from Earth in 1975. All four spacecraft operated successfully. The two orbiters provided global imaging and thermal and water vapor measurements for more than a martian year. The two landers landed on opposite sides of the planet (in July and September, 1976), and obtained measurements from imaging, inorganic chemical, meteorology, seismology, physical and magnetic properties, mass spectroscopic, and biology experiments. Although the other instruments provided a tremendous wealth of information that allowed us to learn about martian geology, the mass spectrometer and the three biology experiments are of the most relevance here.

The biology instrument package consisted of a single mechanism designed to obtain samples and to distribute them to three separate experiments. The three experiments were a carbon assimilation experiment, a labeled release experiment, and a gas exchange experiment.

The carbon assimilation experiment (also referred to as the pyrolitic release experiment) was designed to test for martian organisms that could take in CO_2 or CO from their environment and incorporate them into organic material. This experiment was done under conditions most like those of the ambient martian environment, in order to look for biological activity that could occur in the relatively harsh conditions that occur globally. In the experiment, a sample of soil was exposed to CO_2 and CO gas, either with or without water being present. The gas had been brought from Earth, and was 'labeled' with the radioactive isotope ^{14}C. After a period of time during which metabolism might occur, the soil sample was heated; this would break apart any organic compounds that might have formed and release them as CO_2 (a process known as 'pyrolysis'). Thus, if any CO_2 or CO had been incorporated into material in the soil, it would be released and would show itself by counters that could detect the decay of the ^{14}C isotope. This experiment was designed to test for photosynthesis, chemosynthesis, or other processes of metabolism that could involve CO or CO_2.

The experiment showed that labeled carbon was, in fact, incorporated into the soil, exactly as might be expected if the soil contained active biota. The level of activity, however, was substantially lower than was expected based on experience with terrestrial soils. Further, heating the sample as high as 175 °C for three hours prior to exposing it to the gases reduced but did not eliminate the positive response. Any biological organisms within the sample would have been killed by temperatures that high; even terres-

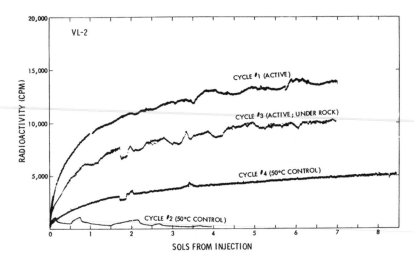

[Figure 8.14]

Data from the *Viking* Labeled Release experiment. The increase in measured radioactive counts with time following the injection of nutrients containing labeled carbon indicates the release of carbon with time. Although the experimenters interpreted the results as being indicative of martian life, the best interpretation of the data from all of the experiments together is that no life was present. (From Levin and Straat, 1977.)

trial biota from hydrothermal systems are not expected to live at such a high temperature in a Mars-like atmosphere. The conclusion reached was that some geochemical process was taking CO or CO_2 from the atmosphere and producing small amounts of carbonate mineral or carbon polymer within the soil. There was no convincing indication from this experiment that biological activity was occurring.

The second experiment was the labeled release experiment. This experiment tested for the presence of martian organisms able to assimilate organic compounds from their environment and, in doing so, release gas back into the atmosphere. Here, organic nutrients labeled with radioactive isotopes were put into contact with the soil. If the nutrients were utilized by organisms within the soil, then any gases given off by the soil would contain some of the labeled atoms. These gases were then tested for radioactivity from the nutrients. The nutrients consisted of molecules of formate, lactate, glucose, glycine, and sulfate in water, and these were labeled with radioactive ^{14}C and ^{35}S; because of the unknown 'handedness' of any martian biota (see Chapter 6), both left- and right-handed molecules were included.

A positive result from this experiment would be a signal of radioactive counts in the gas surrounding the soil upon the addition of the nutrients; this signal would increase with time and then would gradually level off as the nutrients were used up. In fact, that was exactly the signal that was observed (Figure 8.14). Further, heating the sample to 50 °C substantially reduced the signal, and when the sample was heated to 160 °C its ability to produce a positive signal was destroyed entirely. Again, this is exactly the type of signature that would be indicative of the presence of martian biota. However, it is also possible that it represents the response to chemical reactions taking place within the soil. Active oxidants in the soil, such as hydrogen peroxide again, could react with the nutrients to produce the observed signal; interaction with clay minerals in the soil, or with the mineral limonite, could also catalyze the destruction of the nutrient. Although this experiment yielded a positive signal for the detection of life, the signal was not necessarily a unique indication of the presence of life.

The third experiment was the gas exchange experiment. Again, soil was exposed to a mixture of nutrients. The gas surrounding the soil was sampled subsequently, in order to look for specific gases that might be given off by martian biota; the gases tested for were H_2, N_2, O_2, CH_4, CO_2, argon, and krypton, the last two as calibration test gases. The nutrients included a wide range of amino acids, salts, vitamins, CHO compounds, and purines and pyrimidines (the bases in DNA molecules), again with both left- and right-handed molecules. This experiment could be run either in 'wet' mode, with the nutrient solution added directly into the soil, or in 'humid' mode, with the nutrient allowed to evaporate into the space above the soil.

Significantly, oxygen was given off immediately after exposing the soil to the nutrient. However, this was not thought to be a biological response for several reasons. First, the oxygen was given off in the dark, which is different from how terrestrial organisms would be expected to behave. Second, the reaction was not stopped by heating the sample to 145 °C. Third, the same gases were given off (albeit at a different rate) even when the soil was only exposed to water vapor. The results could best be explained, again, by the presence of oxidants in the soil; the water liquid or vapor would displace molecules of other gases that had attached themselves onto the soil grains and would release these other molecules into the atmosphere.

The final relevant experiment was not a formal part of the biology package, but was nonetheless instrumental in properly understanding the implications of those experiments. This was the gas chromatograph/mass spectrometer experiment. It was designed to measure the abundance of organic molecules in the soil, by heating the soil until any complex organics would be decomposed into small organic molecules, evaporated, and then detected. The detection mechanism consisted of first passing the molecules through a gas chromatograph, which is a porous column through which gas can diffuse. With different gases diffusing at different rates depending on their masses, passing the gas through a chromatograph is a simple way to separate gases from one another. The gas that came out of the column was passed into a mass spectrometer, which could determine the abundance of different organic molecules by measuring their masses. The only organic molecules that this experiment detected were various solvents that were left over from the cleaning of the instrument before launch. The upper limit determined for the abundance of organics in the soil was typically at the parts per billion level; this is well below the level expected if there were any active or even dead biota present. Further, a higher level of organic molecules should have been measured even if the only source of organics at the martian surface was from incoming meteorites and asteroids. This experiment placed a very stringent limit on the occurrence of biological activity and suggested the presence of some type of oxidizing molecule in the soil that could destroy organic molecules.

Interestingly, all three biology experiments initially showed the results that would be expected if the soil samples contained active biota. However, the complete absence of organic molecules makes it very difficult to believe that the results could be due to life. The inability of heat to kill the positive response seen in two of the biology experiments supports this view. This points to two important aspects of the *Viking* biology experiments (and, really, of any tests for biological activity). First, it is important that control experiments be performed on the same soil samples that are being tested. Without them, the *Viking* experiments would have led to the conclusion that there was life on Mars. Yet, having done the controls and obtained many of the same experimental results, it is hard to believe that life is the best explanation for the observations. Second, it is important that multiple experiments be performed to test multiple hypotheses. Again, two of the *Viking* experiments initially gave positive results, but the complete ensemble of experiments provided a much clearer picture than did any single experiment.

Searching for life on Mars

The strongest conclusion that can be reached from the *Viking* experiments is that there is no strong indication of the presence of living biota at the *Viking* lander sites. Ultimately, however, the *Viking* experiments were able to test for only a couple of the possible mechanisms by which putative martian organisms might obtain energy. These involved the utilization of either carbon dioxide or certain organic nutrients as a source of carbon in

the production of organic molecules. The nutrients were supplied in abundances far exceeding those that might be found in the ambient martian
environment, and may have overwhelmed any martian organisms.
Martian biota conceivably might metabolize other substances to obtain
energy, and might do so under physical conditions quite different from
those of the *Viking* biology experiments.

Furthermore, if life exists only in isolated oases where liquid water
exists, then the *Viking* experiments might have been the right ones but at
the wrong location. The best places to search for extant life on Mars might
be at the sites of the most recent presence of liquid water at or near the
surface. These would include sites of recent volcanic activity, hydrothermal or hot-spring activity, or catastrophic flood activity. The fact that
active volcanism is likely to be continuing up to the present, and that
hydrothermal systems should be associated with them (possibly including
hot springs that discharge water at the surface), shows that environments
almost certainly exist at the present that are suitable locations for life to
exist. It is also conceivable that life could be found that is associated with
the presence of ice or brines in the regolith. There is abundant evidence for
an ice-rich permafrost, and salt-rich water could exist as a transient phase;
terrestrial life has been shown to exist in such environments, and possible
martian organisms could do so as well.

Life has been shown to exist within rocks in Antarctica (see Chapter 5),
living off of water that can melt within the pore spaces in the rock.
Sandstone is an excellent possibility for such activity, given that water can
diffuse into the rock, crystallize as ice, and then melt on the warmest,
sunniest days. However, the martian environment is substantially colder
than the Antarctic summers, and the possibility of ice or liquid being stable
within a rock has not been demonstrated.

Even if there is no life at the present on Mars, it still is possible that life
existed during some earlier epoch. Life could have formed either during an
earlier, more clement period of martian history or, at almost any epoch, in
near-surface or subsurface liquid-water environments such as hot springs.
As Mars became colder and drier, then, life might have gone extinct. Places
to search for evidence of extinct life would be locations where water had
been abundant, especially during the earlier epochs. These locales might
include the bottoms of the dendritic valleys, where liquid water flowed
prior to about 3.5 b.y.a.; ancient basins, into which liquid water would have
flowed during these early epochs and where liquid might have survived for
long periods of time as ice-covered lakes; and sites of ancient volcanism,
especially where there is geologic evidence for liquid water having been
associated with them at the surface.

An intriguing place to look might be in the walls of Valles Marineris, the large canyon system on Mars. Valles Marineris ('Mariner Valleys', named after the *Mariner* spacecraft that explored Mars in the 1960s and 1970s) extends over almost 3000 km length and is up to 10 km deep in places. It is associated with tectonic features connected to some of the large volcanoes, and shows evidence of both tectonic and water-related processes having occurred there. Where exposed, the walls of Valles Marineris show the presence of layering that may be indicative of the deposition of material in a water-rich environment. In addition, the bottoms of the walls are sufficiently deep that they might show evidence of the liquid water that would have been present several kilometers beneath the surface. Because the valley appears to have formed in part by tectonic processes that left steep-sided walls, the exposed wall regions allow easy access to places that used to be buried many kilometers below the surface, and the possibility exists for exploration of the entire column of crust from the surface down to about 10 km depth.

We can use our estimates of the amounts of energy available in the martian environment to estimate how much life, if any exists on Mars, might have been constructed throughout time. The most likely sources of energy to power martian life would be geochemical energy either from hydrothermal systems (as on Earth, as discussed in Chapter 5) or from chemical weathering of minerals at or near the surface. Both sources of energy are used by various organisms on Earth, and their ready availability on Mars makes them the most likely source of energy for early life. The estimates of the total amount of volcanism on Mars indicate that there was about 1% as much as on the Earth (Figure 7.3). Similarly, geological estimates suggest that the equivalent of a global layer of material perhaps 10 m thick has been weathered at the surface. Based on the amounts of chemical energy available in hydrothermal systems and in the oxidation of surface minerals, it is possible that, over 4 b.y. on Mars, enough life could have been created to make a global layer about 10–20 cm thick. Although this may seem like a lot of life, it is as much as could have been created on the early Earth in only 50–100 million years. Clearly, martian life, if it existed, must not have been very abundant.

How would we go about searching for evidence of existing or extinct life on Mars? First, we would need to identify specific sites that might be suitable for biological activity. Certainly, we have no conclusive evidence that many or any of the possible oases for present-day life exist; candidate locations would need to be identified. These sites could be investigated in detail by lander spacecraft, which would examine their chemistry and geology and look for evidence of organic molecules. They could look for fossil evidence for life, similar to that which has been found in 3.5 b.y. old **141** rocks on the Earth. Ultimately, most scientists believe that returning the samples to the Earth will be necessary in order to prove definitively that life is or was present; of course, the samples need to be chosen from the carefully selected sites, rather than from randomly selected locations on the surface.

Concluding comments

In summary, the available evidence does not provide a unique and compelling indication one way or the other about the existence of life on Mars. The *Viking* landers did not find evidence of life at two specific locations on Mars, and the dearth or absence of organic molecules supports this conclusion. However, it still is possible that life could exist at the present in isolated, more favorable environments on the martian surface. Conditions on Mars may have been conducive to the origin of life, either during early, more-clement conditions or, in hydrothermal systems and hot springs, at almost any point throughout geologic time. It is possible, therefore, that evidence for extinct life exists and is possibly widespread on Mars; no search for such extinct life has been made. Finally, life is likely to have been imported to Mars from the Earth, due to large impacts ejecting material to space; if life was transported to Mars, then it might by chance have found the occasional oasis in which it could thrive.

Mars is the most likely place in our solar system, other than the Earth, to find evidence of extraterrestrial life. Future spacecraft missions are required in order to investigate these possibilities, and a detailed exploration program is necessary. Despite the possible finding of fossil life in a meteorite from Mars (see the next chapter), life on Mars almost certainly will not be found in the first samples that we bring back to the Earth.

142   On August 7, 1996, a press conference was held at NASA Headquarters in Washington, D. C., to announce the discovery of evidence of possible fossil life in one of the meteorites that had come from Mars and been found on the Earth. Although the evidence for life is ambiguous and subject to considerable uncertainty, it provoked a tremendous response from the press and the public, and it triggered a remarkable debate within the scientific community. Clearly, the question of possible life on Mars is of immense interest to everybody, scientists and public alike.

This meteorite, known as ALH84001, is a 1.9-kg rock which is one of the 12 meteorites found on Earth that are thought to have come from Mars (Figure 9.1; see Chapter 8 for a discussion of why we think these are from Mars). It had been found in the Allan Hills region in Antarctica (hence the 'ALH' part of its name) in 1984 (hence the '84') during part of a National Science Foundation expedition that was searching for meteorites. Annual searches for meteorites have been made in Antarctica since the mid-1970s. The meteorites fall on the vast ice sheets (and, of course, everywhere else on Earth, too). The ice flows under the pressure of its own weight toward the ocean, and the movement tends to concentrate the meteorites in places where the ice flows up against the mountains. There are very few terrestrial rocks that accumulate on these ice sheets, making the meteorites easier to identify (Figure 9.2). Over 16 000 meteorites have been collected there by American and Japanese expeditions, and the Antarctic meteorites are now the largest source of meteorites in scientific collections.

When ALH84001 was found, it was classified as a 'diogenite' meteorite, one of a particular class that is thought to have originated within the asteroid belt between Mars and Jupiter. The diogenites are a type of meteorite

[Figure 9.1]
Photograph of the martian meteorite ALH84001. The visible face has been sawed to expose the interior. (NASA photo.)

[Figure 9.2]
Collecting a meteorite off of the ice
during an expedition to Antarctica.
This photo was taken during the
1996 collecting season;
unfortunately, no image was taken
of ALH84001 during its collection.
(Photo courtesy of K. McDonald and
the Chronicle of Higher Education.)

that shows evidence for magmatic or volcanic processes during their earliest history. In 1984, the scientific community had not yet been convinced of a martian origin for any meteorites, and it wasn't until over a decade later that a reanalysis of the diogenite meteorites turned it up as a candidate martian rock. Detailed analysis, particularly involving the oxygen isotope composition (as discussed in Chapter 8), then confirmed that it belonged to the martian clan.

ALH84001 immediately stood out as being distinct from the other martian meteorites, even though it also is one of them. The other 11 meteorites all are young basaltic rocks – volcanic rocks that solidified from molten rock less then 1.3 b.y.a. In fact, it had been the young age of those rocks that first led to the suggestion that they might come from Mars – only the bigger terrestrial planets could retain sufficient heat to produce a volcanic rock this long after planetary formation. ALH84001, on the other hand, was not a basalt, but consisted of essentially pure pyroxene ($(Mg,Fe)SiO_3$), which would have been one of the first materials to solidify from an originally molten planet. And, it had crystallized about 4.5 b.y.a., at the same time that the planets were forming. If it was martian, and if it actually dated from that time, it would be the first meteorite to come from the ancient, heavily cratered martian highlands; on this basis alone, the rock would have merited considerable attention.

The strongest connection to the other martian rocks was through the oxygen isotopes – atoms of oxygen incorporated into the minerals that contained an extra one or two neutrons in the atomic nucleus and, therefore, were heavier than the standard form of oxygen. All 12 of the martian rocks have ratios of the abundances of the oxygen isotopes that connect them to each other and that separate them distinctly from other meteorites, from terrestrial rocks, and from lunar rocks. ALH84001 clearly belonged with the martian rocks. Recall from Chapter 8 that two of the 12 martian meteorites contain gas trapped within them that is similar to the martian atmosphere and distinct from other sources of gas. The trapped gases tie those two rocks, Zagami and EETA79001 (the latter also from Antarctica, the former a meteorite that fell in Nigeria in 1962), closely to Mars. The other ten meteorites are tied to these two, and therefore to Mars, by the comparison of the oxygen isotopes. Although we cannot be absolutely certain that ALH84001 came from Mars, there is a very high probability that it did.

The other characteristic of ALH84001 that sets it apart from the rest of the martian clan is that visible veins of carbonate minerals run through it. These minerals, such as calcite ($CaCO_3$), appear to have been precipitated within the rock from water that was present within it. Although similar features are common in terrestrial rocks, their presence in a martian rock is significant. While some of the other martian meteorites have carbonate minerals, they are present only in small, trace amounts. ALH84001 is the only one that has these carbonate minerals in such high abundances. And, because life on Earth often is associated with environments containing warm water, ALH84001 immediately became a candidate for a search for evidence of biological activity. The fact that the rock dated from a period on Mars in which liquid water might have been stable at or near the surface (Chapter 8) contributed to the interest in this rock. David McKay and Everett Gibson, both scientists at NASA's Johnson Space Center in Houston, along with their colleagues, began a detailed search for evidence that might pertain to life. It was this search that culminated in the 1996 press conference.

ALH84001

Our current understanding of the history of ALH84001 suggests that it has had a moderately complicated history (Figure 9.3). It originally crystallized about 4.5 b.y.a., during the same time period in which Mars was originally forming. About 4.0 b.y.a., it was shocked by some substantial event on the surface of Mars. That event probably was the impact of a large planetesimal onto the martian surface during the tail end of the accretion of the planet. While the impact was strong enough to partly 'reset' one of the radioisotopic clocks and to produce some shock damage to the rock, it did not eject the rock from the surface to space. Sometime later, probably between about 3.6 and 1.3 b.y.a., the carbonate minerals were deposited within the rock. Unfortunately, the age of these carbonates is poorly known at present. Again, the rock was still on (or beneath) the martian surface at this time. The evidence for possible martian life is contained within this carbonate material, so if they are indicators of the presence of life then the life must have existed during this middle epoch of martian history.

[Figure 9.3]
Events in the history of the martian meteorite ALH84001. The occurrence of each of the geologic events is indicated by geochemical evidence, and the dates are determined using various isotopic indicators.

4.5 b.y.a.	Crystallized from molten rock
4.0 b.y.a.	Shocked by an impact, but not ejected from surface
1.8–3.6 (?) b.y.a.	Carbonate minerals deposited by water flowing through the rock
16 m.y.a.	Ejected to space by an asteroid impact
1984	Collected in Antarctica and classified as a diogenite meteorite
1994	Reclassified as a martian meteorite

About 16 million years ago, the rock was ejected into space, presumably by the impact of another asteroid onto the surface. The orbit that it followed around the Sun after being knocked into space must have been complicated, and a number of close passes by Mars would have perturbed the orbit significantly. Its orbit evolved until it eventually crossed the orbit of the Earth. Thirteen thousand years ago, after about 16 million years in space, it landed in Antarctica.

The dates for each of these events were derived from quantitative measurements of radioactive isotopes within the rock – either those that decay through time or those that are created in space by the interactions of solar and galactic cosmic rays with the rock. Of all the facts regarding the history of ALH84001, these events and the dates associated with them probably are the best understood and most universally accepted.

The evidence of possible life contained within the carbonates takes on several forms. None of them is the so-called 'smoking gun' that would point uniquely and exclusively to biological activity, and each can have nonbiological explanations. The original evidence that was presented at the 1996 press conference includes the presence of compositional layering within the carbonates that is similar to layering in terrestrial carbonates and can be produced by biological activity; the presence of mineral grains that, on Earth, are associated with certain types of bacteria; the presence of organic molecules that may have been produced by the decay of living matter; and the presence of sausage- or rice-shaped objects that might be fossil martian bacteria. Each of these will be described in more detail below, along with some of the alternative explanations for their occurrence. In addition, the tremendous interest in the question of martian life has fostered a wide variety of additional observations and measurements within these meteorites, and some of these will be discussed as well.

First, however, is it possible that any of these features might be due to contamination in the laboratory or from the meteorite having resided within ice fields in Antarctica for 13 000 years before collection? The short answer is 'probably not'. The meteorite handling facility at the Johnson Space Center has been dealing with lunar and meteorite samples for almost three decades, and has developed very tightly controlled procedures that minimize any possible contamination. The samples are stored in airtight containers in cleanrooms. Although terrestrial bacteria certainly are not completely stopped by these precautions, their effects on the samples certainly are limited.

During analysis, the samples were handled in ways that minimize possible contamination. They were moved using dust-free gloves, and sample preparation was done on a clean aluminum surface. In order to demonstrate that there would be minimal contamination in this environment, the scientists exposed clean quartz disks to the same procedures and analyses as the rocks; these disks showed no terrestrial contamination, suggesting that the 'fossils' were not introduced by the laboratory environment. To minimize further any effects of contamination, the rocks were cut in a clean environment, and only portions that had been within the interior of the rock were analyzed. In fact, the organic molecules that were found showed a higher abundance deeper within the rock, suggesting that they could not have diffused into the rock from the work surface.

When the rocks were collected in Antarctica, they were never touched by human hands. Rather, they were picked up using sterile, clean bags, and sealed in airtight containers until they were examined back in the cleanrooms at the Johnson Space Center.

Of course, the meteorite could have been altered during the 13 000 years that it spent on the ice in Antarctica prior to being collected. Contaminants in the ice could have migrated into the meteorites, or exchange of atoms or molecules with the ice could have occurred. This possibility was tested by doing the same analyses on other meteorites that also had been collected in Antarctica as were done on ALH84001. These meteorites had gone through a similar history of exposure to the terrestrial environment prior to their collection. If there had been any contamination at that time, it might be expected to show up in these rocks as well. They showed no similar features, however.

It is most likely that the features that were seen within the rocks did not come from terrestrial contamination. Rather, they represent materials that were incorporated into the rock while it was still on the martian surface. Although there clearly is considerable uncertainty as to whether they represent fossil martian biological activity, they just as clearly must represent the results of processes that had occurred on Mars.

Let's get to the specific evidence that was found within the meteorite:

There is compositional layering within the carbonates that may be biogenically produced. Individual carbonate 'globules' are typically about 50 microns across (0.05 mm) (Figure 9.4). Their composition has been investigated using a form of high-resolution electron microscopy, with state-of-the-art equipment, that allows them to be studied in detail. The chemical composition of the globules shows a clear, layered structure (Figure 9.4). Around the edge, there is distinct compositional variation, with alternating magnesium-, iron-, and calcium-rich carbonates. The iron-rich layers, for example, consist mainly of small grains of the mineral magnetite (Fe_3O_4), with some pyrrhotite (FeS) mixed in. This type of layering can result on the Earth when the precipitation of the minerals is aided by bac-

148

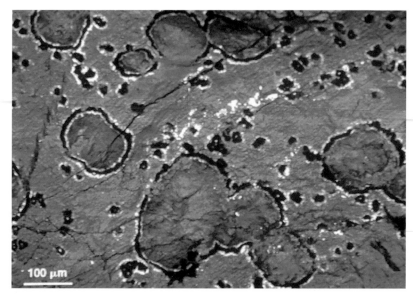

[Figure 9.4]
Photomicrograph showing a thin section through the ALH84001 martian meteorite. The nearly circular objects are carbonate deposits filling voids within the rock. The apparent layering around the rim consists of carbonates with different amounts of iron, magnesium, and calcium. (NASA photo, after McKay *et al.*, 1996.)

teria. The center of one globule contained another iron-sulfide mineral grain as well, although the specific mineral could not be identified.

In particular, the pyrrhotite and magnetite grains may suggest the presence of bacteria. The presence of nonbiologically produced carbonate grains, magnetite grains, and pyrrhotite grains, all next to each other, is thought by some to be puzzling. These minerals would not be expected to have formed under the same geochemical conditions – the conditions under which the last two would form may cause the carbonates to dissolve. However, living organisms produce conditions that are very much out of chemical equilibrium, and could result in the production of all of these minerals in the same location. In fact, precipitation of iron-sulfide and magnetite grains can occur within individual bacteria. And, the iron-sulfide mineral grains are very similar in size and shape to terrestrial 'magnetofossils' – magnetic mineral grains that occur within fossil bacteria and which are clearly connected to biological activity.

A second possible martian 'fossil' found in the meteorite is a particular type of organic molecule called polycyclic aromatic hydrocarbons, or PAHs. PAHs consist primarily of multiple carbon rings bound to each other. The carbon rings themselves are composed of six carbon atoms bonded together into a hexagon shape. These rings are attached to each other, forming structures that look like interlaced patio tiles (Figure 9.5); bound onto some of the carbon atoms are occasional hydrogen atoms or groups of atoms containing hydrogen. On Earth, PAHs are often found as a chemical byproduct of the combustion of organic molecules (such as hydrocarbons

[Figure 9.5]
Schematic diagram showing the
structure of several polycyclic
aromatic hydrocarbons. The
complete structure is shown for
naphthalene, along with a simpler
schematic diagram in which each H
atom is not shown. The total atomic
mass of each PAH is indicated in
parentheses. Naphthalene and
methyl naphthalene are found
in some meteorites. The others
have been found in the martian
meteorites. (After Wing and
Bada, 1992.)

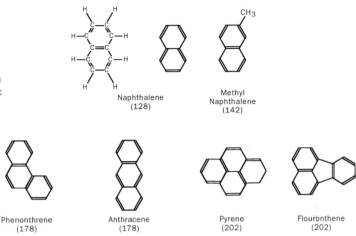

in gasoline) or as the decay products of biological material. The hydrogen
and oxygen are driven off in combustion, leaving the carbon behind in a
complex structure. As such, they are the remnants of the biological activity
that created the original organic molecules from which they formed.

The PAHs are present in the meteorite at a concentration of about one
part per million. This is well above the part-per-billion level of recent
snowfall in Antarctica that contains terrestrial, industrially produced
molecules. The PAHs that are seen have, typically, three to six carbon rings
attached together. These show distinctive masses at 178, 202, 228, and 252
atomic mass units (a.m.u., equivalent to the total number of protons and
neutrons in the molecule) (Figure 9.6). These masses are characteristic of

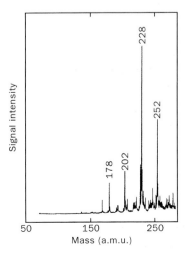

[Figure 9.6]
Mass spectrum of organic
molecules found in ALH84001. The
peaks at 178, 202, 228, and 252
a.m.u. correspond to the masses of
PAHs. In addition, mass peaks show
up at larger masses (not shown).
(After McKay *et al.*, 1996.)

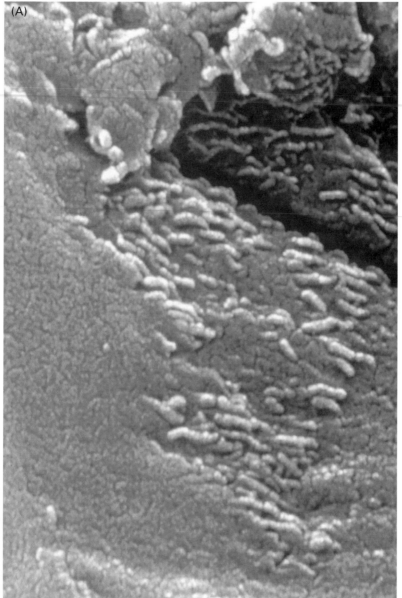

(A)

[Figure 9.7 (A) and (B)]
Examples of possible fossil martian bacteria. (NASA photos.)

the molecules phenanthrene ($C_{14}H_{10}$, with a mass of 178 a.m.u.), pyrene ($C_{16}H_{10}$, 202 a.m.u.), chrysene ($C_{18}H_{12}$, 228 a.m.u.), and either perylene or benzopyrene (both with the formula $C_{20}H_{12}$, 252 a.m.u.). There is evidence for a number of additional, heavier PAHs, with masses from about 270 a.m.u. up to 450 a.m.u.; these are much less distinct and much less abun-

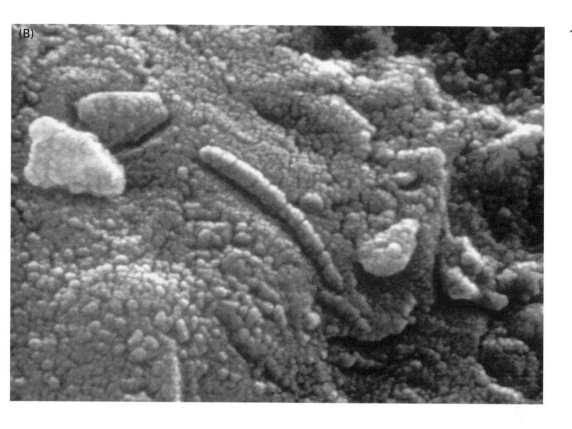

(B)

dant, however, and cannot be identified readily. Although PAHs may be produced by nonbiological processes and are, in fact, found in meteorites that aren't from Mars, they also may be produced by biological processes; as on Earth, these would involve the decay of preexisting biota.

Finally, we have arguably the most intriguing fossil evidence for possible martian life – the presence of sausage-shaped structures that might be the fossil bacteria themselves. These are seen partially embedded in or sitting on top of the carbonate grains, and would have to have been deposited at the same time as the carbonates themselves. They are typically up to about 100 nm long and, in some cases, only 10–20 nm wide (Figure 9.7). This size is equivalent to 0.1 microns, or 0.0001 mm, so these are incredibly small features. Although they are much smaller than the smallest terrestrial bacterial fossils, which are about 100 times larger, they are comparable in size to features that are seen in terrestrial rocks and that have been called 'nanobacteria' (Figure 9.8). Terrestrial nanobacteria have been found as fossils within a number of calcite deposits, and are thought to be biological despite a tremendous uncertainty as to how such a small

152

(A)

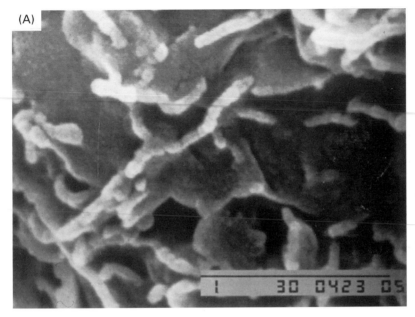

[Figure 9.8 (A) and (B)] Examples of terrestrial nanobacteria. (A) Scoria (volcanic debris) sample from Vulcano, Sicily; scale bar is 1 micron. (Photo by R.L. Folk and F.L. Lynch.) (B) Sample from hot springs at Viterbo, Italy. (Photo by R.L. Folk.)

biological entity could function; however, there is not even any universal agreement that these terrestrial features are biological.

The strongest evidence pointing toward the possibility of these fossils originating from bacteria-like entities is their appearance. They are very similar to bacteria, despite the difference in size, and generally are unlike non-biological geochemical structures. This point was brought home by a story told by Everett Gibson, one of the scientists involved in the analyses. He took a microphotograph of one of the possible bacteria home and showed it to his spouse, who is a microbiologist. She took one look at it and asked what kind of a bacterium it was!

Is biology the only explanation?

Although each of the characteristics described above can be ascribed to biological activity, this is not a unique interpretation. In fact, we have to be careful to discuss alternative explanations before we can decide whether a biological or a nonbiological process is the most likely culprit in their production. Some of the issues and alternatives will be discussed below.

One of the key issues in understanding whether these rocks could

(B)

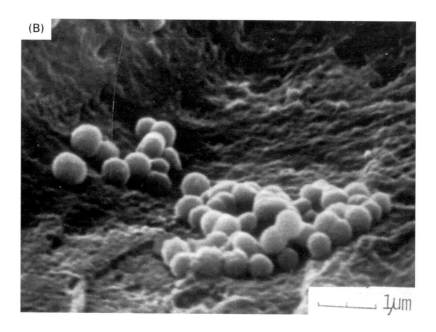

contain fossil martian life is the temperature at which the carbonate minerals were deposited within the veins in the rock. It is within these carbonate minerals that all of the possible biological fossils are found. If the carbonates were deposited at high temperatures like those found in hydrothermal systems, say at temperatures as high as 700 °C, then a biological origin for these features can be ruled out – organic molecules and living entities would be destroyed by the heat. A deposition at much lower temperatures, while not proving that these features are biological, certainly would be much more consistent with that possibility.

Determining the temperature of carbonate formation has become a hot issue (so to speak). Each new analysis since the August press conference has made national headlines in the newspapers. How can we determine the temperature at which these carbonate minerals were deposited? There are several methods that can provide some information. In the first approach, the composition of the carbonate minerals in the globules may be explained by the precipitation of minerals in water at a temperature of around 680 °C. At this high temperature, the combination of minerals seen in the layers could be produced by hot, CO_2-rich water flowing through a silicate rock that had not previously been altered chemically. The existence of discrete layers could result from the passage of a 'pulse' of hot water passing through the rock, rather than a long period of steady hydrothermal activity. If cooling occurred while the minerals were still precipitating, then a layered compositional structure might result. The pulse of heat could come from a nearby impact event, rather than from a steady process involving volcanism and subsurface heating. Unfortunately, this analysis does not lead to a unique temperature determination, partly because the minerals could be metastable at lower temperatures (so that the estimated temperature of 680 °C is an upper limit on the temperature rather than a true estimate).

The second approach to determining the temperature of carbonate formation involves analysis of the ratios of the isotopes of carbon and oxygen within the carbonates. The precipitation of carbonate minerals from water will preferentially take the lighter isotope of oxygen. The degree to which the lighter isotope is preferred depends on the temperature – at higher temperatures, this preference is smaller. Therefore, if we know the initial ratio of the oxygen isotopes in the system, the ratio in the carbonates can be used to derive the temperature at which they deposited. This technique is a standard method of deriving the temperatures at which minerals deposited in terrestrial groundwater systems, and generally is accepted as accurate. Applied to the carbonates in ALH84001, it results in an estimate of the water temperature of between about 0 and 80 °C. At these temperatures, of course, life could exist quite readily.

There are a couple of problems with the isotopic estimates, however. First, the carbon isotope ratio does not appear to be consistent with this interpretation. However, the interpretation of the carbon isotopes depends strongly upon how much carbon is available within the martian volatile system, and upon the likelihood of the carbonate formation depleting the remaining carbon reservoir. Second, the martian oxygen and carbon is not a closed system. Both species can be lost to space over geologic time, with loss preferentially removing the lighter isotope. In essence, this means that the starting point for the isotope ratios is not known. If there has been substantial loss to space, and if this has affected the initial oxygen and carbon isotope ratios prior to carbonate formation, then the result would be an estimated temperature for the water higher than the 0–80 °C values. The temperatures could have been as high as about 250 °C and still be consistent with the isotope data.

A third way of determining the temperature comes from the ability of certain magnetic minerals to retain some memory of the magnetic field in which they formed. If the temperature of the mineral is raised to high values, this memory is lost. Measurements of the magnetic field retained by two magnetite grains within the rock suggests that the temperature has not been above about 300 °C since the time when the carbonates were formed. Interestingly, this analysis also suggests that Mars had a sub-

stantial magnetic field at the time that the mineral grains crystallized;   **155**
today, it has either a very small field or no field.

There also are alternative explanations for some of the other characteristics observed in ALH84001. In particular, let us address the PAHs and the
things that look like fossil bacteria in the meteorite.

PAHs are common on Earth and, in fact, in non-martian meteorites as
well. In meteorites, they are thought to form from CO_2 and H_2O that were
present in the interstellar cloud that preceded the formation of the solar
system; they later were incorporated into the material that became the
meteorites. The process by which PAHs formed in the solar nebula is not
well understood, but probably involved a series of chemical reactions
between the various gases in the cloud. These reactions would have taken
place on dust grains that could act as a catalyst, helping drive reactions
that otherwise would proceed very slowly. Primitive meteorites have other
organic molecules within them, in addition to the PAHs, including such
complicated molecules as amino acids. As discussed in Chapter 6, these
meteorites may have been a source of organic molecules on the prebiotic
Earth.

The PAHs on Mars probably were not formed by the same process as in
the meteorites, and they probably did not arrive along with the material
that formed the planet. The processes that formed the PAHs within the
interstellar medium probably would not have operated in the martian
environment. Although different processes could operate, they would best
occur under the conditions that have already been discussed for production of organic molecules (see Chapter 6). If the PAHs arrived on accreting
meteorites and planetesimals, it seems unlikely that they would be able to
survive the active geological environment that would have been operating
on early Mars. The energy from impacts would readily destroy PAHs during
this early epoch, and geological processes such as volcanism would do so
later on. In addition, the martian surface today is a very oxidizing environment in which organic molecules quickly would be destroyed (see Chapter
8). It seems unlikely that PAHs from early martian epochs could survive
long enough to have been emplaced in the rock so long after the
crystallization of the rock.

Does this mean that the PAHs are biological in origin? While they
indeed are organic, they may not be biological. They may have formed
from inorganic processes involving chemical reactions between preexisting organic molecules. In the same way that they could form on the
Earth from hydrocarbons, PAHs may have formed on Mars from nonbiological, partially reduced, carbon-containing molecules such as amino
acids or even methane; oxidation of these molecules would remove the

hydrogen and oxygen and leave behind the reduced carbon. Regardless of whether they formed biologically, the PAHs clearly demonstrate the existence of organic molecules in the martian environment. Because an origin of life on Mars requires the presence of organic molecules, the presence of martian organics has substantial implications for the possible formation or existence of life on Mars; this is the case even if the PAHs did not form directly from living organisms.

How do the possible martian bacterial fossils compare with terrestrial biota? Most importantly, they are substantially smaller than the oldest known terrestrial bacteria. The oldest terrestrial bacterial fossils are 3.5 b.y. old, and appear identical in size, shape, and morphology to modern-day cyanobacteria that are seen in stromatolite deposits (Chapter 4). There are two important comparisons to make here. First, it would be difficult to try to construct a martian bacterium of the observed size from terrestrial components. The cell wall alone in a terrestrial bacterium is about 25 nm thick; if martian biota used the same structure for support, there would be no room left for anything else to exist inside the walls (Figure 9.9). Although this may seem like a powerful argument against these fossils being biological, it is not fatal – there is nothing that requires martian bacteria to have found the same solution to containing their insides as terrestrial bacteria.

Interestingly, the martian fossils are about the same size as terrestrial ribosomes (Figure 9.9). In terrestrial cells, ribosomes perform the function of taking the information in messenger RNA and using it to construct proteins. The ribosomes themselves are comprised of a combination of proteins and ribosomal RNA; about 60% of the weight of each is made of rRNA. Ribosomes are only a part of the total machinery involved in reproduction, however, and could not exist independently from the cells in which they reside.

The martian fossils also are about the same size as terrestrial viruses (Figure 9.9). Viruses consist primarily of genetic material (either DNA or

50 nm

Martian fossil

Ribosome

Virus

Membrane

[Figure 9.9]
Comparison of the sizes of the possible martian fossil bacteria, a terrestrial ribosome, a virus, and the smallest membrane that occurs within a cell.

RNA), often contained within a protein shell. Viruses are unable to reproduce by themselves, but attack individual cells (either isolated bacteria or
cells within a larger organism) and use the cell's own reproductive machinery to copy themselves. Because they are incomplete cells, in this sense,
they occupy some boundary between being either the smallest living entities or the largest non-living macromolecules.

Thus, the terrestrial biological structures that are closest in size to the
martian fossil structures are not independent, living cells capable of reproduction and metabolism. However, a volume the size of these martian
objects still is large enough to hold about 1000 DNA base pairs. Although
there is nothing to require that martian organisms would use DNA, it is
clear that they could contain a substantial amount of genetic information.
Is this enough? The fact that terrestrial bacteria use so much more DNA
than these martians could contain suggests that there might be a
minimum amount of genetic information required to function as life. Of
course, even the simplest terrestrial organism is a very advanced organism
both biochemically and genetically, and the oldest known organism certainly must have advanced well beyond the simpler organisms that would
have been the first living terrestrial entities. The first life must have been
much smaller and simpler, possibly being as simple as molecular chains
that reproduced on the surfaces of mineral grains (see Chapter 6). As we do
not know what stage of life these martians represent, it would be pre-
mature to rule out their being alive.

Is there fossil life in ALH84001?

When all of the evidence is considered, does it demonstrate the existence
of a fossil martian biosphere? It should be clear that each of the major
characteristics observed in ALH84001 could have either a biological
explanation or a nonbiological explanation. Even the original researchers
pointed this out in the presentations of their results. They argue, however,
that the concept of 'Occam's Razor' supports a biological origin. This
concept suggests that, when alternative mechanisms are available to
explain a given phenomenon, the simplest explanation or the one that
makes the fewest *ad hoc* assumptions, must be given preference.

The several lines of evidence can be explained either by (i) a single
process – biology – acting in ways that are reasonably well understood, or
(ii) a variety of nonbiological processes acting, with each process being
required by one or more of the observed characteristics. David McKay and

his colleagues argued that biological activity is the simpler explanation because it makes fewer requirements on the environment or the geochemical system; therefore, it should be given preference. On the other hand, others have suggested that, in choosing between inorganic, nonbiological processes that might have operated on Mars and the possibility that biological processes might have operated on Mars, the nonbiological processes automatically are inherently simpler and must be given preference. Even if a multitude of inorganic, nonbiological processes must have acted together, they argue, such behavior would be simpler and more reasonable than the suggestion that life has operated on Mars.

Which is correct? Unfortunately, not enough information is available today to determine which is the better explanation. There is not enough evidence to make biological systems the most plausible explanation, nor is there enough evidence to make inorganic systems the most plausible explanation. More data are needed, both from the martian samples that we have in our collections and from well chosen samples from interesting and relevant places on the martian surface. Certainly, many people will be surprised if the one rock that we have from the ancient surfaces on Mars turns out to contain evidence of life, especially when so few ancient rocks on the Earth contain evidence of terrestrial life.

What would it take to demonstrate the past or present existence of a martian biosphere? With the samples that we have in hand today, it would help to demonstrate the existence of structures analogous to cell walls in the possible bacteria fossils, to show examples that appear to be in the process of replicating (i.e., in the process of dividing, as is seen in many terrestrial examples of fossil bacteria), or to show interior structure that might otherwise demonstrate a biological origin of the original object. The techniques that were used on the martian rocks are state-of-the-art, though, and significant advances will take considerable additional effort. Importantly, there have been no equivalent studies of non-martian meteorites or even of terrestrial rocks. If a detailed investigation of those rocks produced no evidence of similar, nonbiological features, it would suggest that the martian features were more likely to be biological. In samples that may be returned from the martian surface by spacecraft, we would look for essentially the same types of features. In addition, however, we would know the geological context for the samples which would provide us with important additional information for interpreting the observations.

Ultimately, it appears that martian sample return missions will be required in order to determine if there was in the past or is today a martian biosphere. The question of whether there is one has been around in serious scientific circles for at least the past 30 years and arguably even for the past 100 years. It probably will not be answered for another ten or more years.

Concluding comments

There is an important point to keep in mind in the ongoing discussion of the nature of the fossil evidence for possible martian life in the martian meteorites. If detailed analysis demonstrates that these rocks do not contain any evidence of martian life – that all of the features that have been found can best be explained by geological and geochemical processes acting on Mars – the intellectual arguments that suggest that life might occur on Mars (Chapter 8) will not change. There still will be every reason to believe that life could have originated on Mars, either at the surface during the early epochs when water was stable there or later in martian history, buried beneath the surface. The question of the existence of martian life will be answered only by going to the most likely places to contain life and finding out what is there. As discussed in Chapter 8, these might include hot springs, sites of recent volcanism, underground hydro-thermal systems, or deeply buried aquifers.

160   The geological history of Mars was discussed in Chapter 8, including the behavior of volatiles such as CO_2 and H_2O and the possible origin and existence of indigenous martian life forms. There are aspects of life on Mars that go beyond this, however, at least in the broadest possible context of what we mean by life. In particular, there has been a fair amount of discussion about the possibility of implanting life on Mars. This could involve the deliberate introduction into the martian environment of terrestrial organisms, with the intent of having them grow and multiply. In this context, it would require 'terraforming' Mars, whereby the martian environment would be altered into a sufficiently Earth-like environment that terrestrial plants, or even humans, could survive. In addition, it is possible that we could unintentionally introduce terrestrial organisms onto Mars. This could happen as we continue to send spacecraft there, with Earth organisms going along for the ride; conceivably, such hitchhikers could find a suitable environment on Mars where they might survive and multiply. Finally, although usually grouped with the idea of UFOs, some people believe that the so-called 'Face on Mars' could be evidence of a preexisting, advanced civilization that lived on, or at least visited, Mars.

Most of these issues clearly are not at the mainstream of planetary science or exobiology today. However, they are important for understanding the place that Mars holds in our discussion of extraterrestrial life. For this reason, these three aspects of extraterrestrial martian life will be explored in this chapter.

Terraforming Mars

The idea of being able to terraform a planet has become deeply rooted in our culture, at least in the science fiction literature of the last couple of decades and in today's 'Star Trek' and 'Star Wars' cultures. Mars is of substantial interest in this regard because it is the most Earth-like planet in our solar system, has a climate that is already somewhat similar to that of the Earth, and is seen as a plausible, if not likely, target for colonization by future terrestrial astronauts. Is it feasible to consider doing this? What are the technological and scientific issues associated with terraforming Mars? And, finally, what are the moral and ethical issues connected to enacting such a dramatic change to the environment of another planet? These questions have been considered recently by a number of researchers. Much of the discussion here is taken from a series of papers by Christopher McKay of NASA's Ames Research Center in California and his colleagues.

We will consider the requirements that are placed on an altered, terraformed, martian environment by the desire to support life, the degree to which such a martian climate might be stable, how we might engineer the transformation from the present climate to such a terraformed climate, and whether there are sufficient gases available in the martian environment to provide such a changed climate.

If we desire to have terrestrial life survive in a changed martian environment, we need to consider the composition, pressure, and temperature of the atmosphere that would allow life to exist in the new environment. The minimum requirements will be different for a planet that can support terrestrial plants and for a planet that also is capable of supporting humans.

The main constraint on surface temperature is the requirement that liquid water not freeze. This usually is interpreted to mean that the temperature must be above 0 °C. Although there are life forms that can exist at colder temperatures, within permafrost, they are extremely limited and generally do not require an environment that fits into our view of what would constitute a terraformed planet. There are two important points to be made with regard to this requirement. First, as noted in Chapter 8, the temperature on Mars today actually rises above freezing at the equator on many if not most days out of the year. Despite this, the average temperature at the surface is some 50–60 K below freezing, making it hard to imagine that any plants or animals could survive. Therefore, the requirement for the temperature probably is that the average daily temperature be at or above freezing. Then, even if the nighttime temperatures were below freezing, plants might have enough thermal mass or thermal inertia to keep from freezing overnight. Similarly, we also imagine that we are talking about the annual average temperature, where plants might survive a cold, harsh winter by virtue of having had a warm, pleasant summer.

Second, most people will recognize that water on the ground will not necessarily freeze even when the air temperature drops below freezing. This is because the ground will absorb sunlight directly and warm up more than the bottom of the atmosphere. In winter, streams may not freeze until the air temperature has dropped to well below freezing. In Antarctica, where temperatures regularly can average more than 20 degrees below freezing, there are lakes that never freeze completely; they are kept liquid beneath a covering of ice by the steady input of sunlight and by an influx of stream water during the few days each year that temperatures might rise above freezing. Observations suggest that these ice-covered lakes can keep from freezing even at an average annual surface temperature of 245 K, almost 30 K below the freezing temperature of ice.

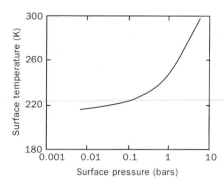

[Figure 10.1]
Model of the greenhouse behavior of the martian atmosphere. The curve shows the surface temperature that will result for a given atmospheric CO_2 pressure. This model assumes that the energy output of the Sun has its current value. (After Kasting, 1991.)

For our considerations, however, the annual average temperature at the bottom of the atmosphere probably needs to be at or above the melting point of ice. These increased temperatures can be produced by the presence of greenhouse gases in the atmosphere. These gases, such as CO_2, H_2O (as vapor), or ozone, inhibit the ability of the atmosphere to radiate heat to space, and thereby allow the atmosphere to warm up. The 6 mbar of CO_2 that is present within the martian atmosphere today provides a greenhouse warming of about 6 K. On Earth, greenhouse warming is provided primarily by CO_2 and H_2O, and it increases the surface temperature by about 30 K above what it would otherwise be. An increase in the martian temperature by 50–60 K would require a CO_2 atmosphere that contained 1–2 bars of gas (Figure 10.1).

The amount and composition of the gases that are necessary to support life functions depend directly on the metabolic requirements of different species. For plants, photosynthesis requires the presence of CO_2. Most terrestrial plants cannot survive on less than about 0.15 mbar of CO_2; there is no clear upper limit for most plants on how much CO_2 can be present. Access to nitrogen is required, with a minimum abundance of 1–10 mbar. Oxygen also is required, even for photosynthesizing plants; they generally require more than about 20 mbar O_2.

Humans are more demanding in their needs. The CO_2 abundance would need to be less than about 10 mbar in order to not reach toxic levels. The O_2 abundance would need to be more than about 130 mbar (compared with the present terrestrial atmospheric value of 210 mbar). Nitrogen gas is not required *per se*, but would need to be present as a 'buffer' gas in order to bring the total pressure up to about 500 mbar; this minimum total pressure is comparable to that on the highest mountains on Earth.

It seems clear that the requirements on pressure, atmospheric composition, and temperature are mutually exclusive for Mars when both plants and humans are considered. A high abundance of CO_2 gas is necessary to bring the temperature up; such a high CO_2 abundance might be conducive to plant survival, but would be deadly for humans. The Earth is able to reach a happy medium primarily because it is closer to the Sun and less greenhouse warming is needed to keep the temperatures above the melting point of ice.

A solution to this problem might be to find an alternative greenhouse gas capable of raising the temperature but without being detrimental to humans. Chlorofluorocarbon compounds (CFCs), such as are used in refrigerators and air conditioners, might be able to do this. The problem with using them, however, would be that they decompose in the presence of sunlight, and might have a lifetime of only several hundred years. Thus, they would need to be continuously manufactured. A rough estimate of the manufacturing rate required to keep the atmospheric abundance of CFCs at a high enough level suggests that more than 10^{12} tons/yr would have to be produced. Comparison with the current terrestrial production rate, about 10^6 tons/yr, suggests that this approach may not be easy to achieve.

An alternative approach might be possible based on an inherent instability in the martian climate system. At the present low atmospheric pressure, there is very little transport of heat through the atmosphere to the polar regions. This allows the poles to become very cold in the winter season, and CO_2 frost will deposit onto them; this condensation serves to keep the total atmospheric pressure relatively low. This is the opposite of the situation on Earth, where enough heat moves poleward into the winter hemisphere through the atmosphere and the oceans to keep the polar oceans from freezing. On Mars, a slight increase in the atmospheric pressure would result in an increase in the amount of heat that would be transported poleward. An increase in pressure of as little as several tens of millibars might increase the poleward heat transport to high enough values to lead to an instability: the extra heat moved into the polar regions would heat up the polar caps and high-latitude regolith, causing CO_2 that is stored there to be released into the atmosphere. The resulting increase in pressure causes a further increase in poleward heat transfer, again triggering the release of more CO_2. This positive-feedback loop might not reach equilibrium again until between 100 mbar and 1 bar of CO_2 resided in the atmosphere, if that much CO_2 is available in the polar regions (Figure 10.2).

It is possible that this climatic instability mechanism may have been responsible for dramatic changes in the martian atmospheric pressure throughout geologic time. The largest uncertainty in whether it could be triggered is whether there is enough CO_2 available and accessible in the martian polar caps and high-latitude regions. The polar caps might

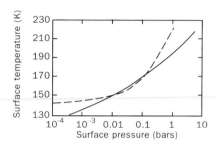

[Figure 10.2]
Pressure instability of the martian atmosphere. The dashed line shows the temperature as a function of the pressure of CO_2 in the atmosphere. The solid line is the saturation temperature of CO_2. Wherever the model predicts a pressure greater than the saturation pressure for a given temperature, the climate system is unstable. In this case, the pressure will drop until it reaches about 5–10 mbar. (After Haberle *et al.*, 1994.)

contain the equivalent of as much as a fraction of a bar of CO_2 as ice. The ground might contain as much as another bar of CO_2. However, diffusion of CO_2 from the regolith into the atmosphere will be inhibited by the ground ice that is expected to be present near the surface (see Chapter 8). In addition, the CO_2 in the ground is in the form of adsorbed gas, held in place by relatively weak physical bonds. It can be released most easily by heating up the regolith; releasing much or all of the gas requires heating up the regolith to a depth of about a kilometer. The only way to do this is to heat the surface and let the heat diffuse into the subsurface; it would take up to 10^5 years for heat to diffuse through a kilometer of regolith. Thus, it is not clear that this instability can be triggered without first warming up the climate for very long time periods.

And, of course, if the CO_2 could be released into the atmosphere and lead to a warmer climate, the warmer temperatures would cause the atmospheric CO_2 to dissolve in any newly formed surface or subsurface liquid water. The dissolved CO_2 then would be able to form carbonate minerals, effectively removing the CO_2 from the atmosphere again. This process would remove the CO_2 from the atmosphere in only 10^7 years. In other words, a CO_2 greenhouse atmosphere could be created, but it would not be stable; the CO_2 will move preferentially into the subsurface.

In place of CO_2 as a greenhouse gas, there are alternative gases that could be used to raise the atmospheric temperature, if one is willing to rely on technology that does not yet exist. For example, nitrogen in the atmosphere could be transformed into ammonia, NH_3, which is a very efficient greenhouse gas. Even the small amount of nitrogen in the present atmosphere would be sufficient to provide enough ammonia to raise the global temperatures to the melting point of water ice; converting this nitrogen to ammonia is a very difficult process, however.

Is terraforming Mars feasible? Right now, it is difficult to say one way or the other with any certainty. It depends on how much CO_2 is tied up in the martian polar caps and regolith, and on the detailed physics of the instability inherent to the martian climate system. Neither of these is known with any certainty. It is interesting to note that, despite the tremendous difficulty that we would have in changing the climate on Mars to one that would allow liquid water, Mars apparently had no difficulty in doing this some 3.5 b.y.a. Despite the fact that the solar luminosity at that time was 30% less than its present value, liquid water appears to have been relatively widespread on early Mars (Chapter 8). Presumably, this was the result of greenhouse gases that were produced naturally and maintained in the martian atmosphere by ongoing geologic processes. The transition to a colder climate presumably was the result of loss of the greenhouse gases to crustal reservoirs or to space.

The moral and ethical dilemma of terraforming Mars

If it turns out to be feasible, should we consider terraforming Mars? This gets into ethical and moral questions that are inherently different from those connected to environmental issues on the Earth. Chris McKay has examined the Earth-bound environmental movement for guidance, and has been able to construct three general approaches as to how the environment can be viewed. These may be applicable to the martian situation.

The first approach suggests that nature has by itself provided the most appropriate environment, and that to tamper with it in any way is inappropriate. In this 'anti-humanism', as he terms it, any human activity that can affect the environment in any way is to be avoided. The second approach is 'wise stewardship', in which attributes of nature can be used by humans, but only in a manner that does not squander their value. This approach is used most often in deciding how to use terrestrial natural resources, although there is often fundamental disagreement as to what constitutes appropriate or wise use. The third approach is one of 'intrinsic worth'. Here, it is recognized that the non-human aspects of nature have intrinsic value and worth, independent of their value to humans, and that those rights should be respected. These three approaches are extreme 'end-members' in the possible views that one can have of the environment; it is easy to imagine an individual view that combines aspects of each approach.

Does Mars as a planet have any intrinsic value in and of itself? Is there less intrinsic worth in a planet that is devoid of life than in one with an active biosphere? Should we access and use the resources that are available there or should we leave them as they are?

166   Robert Haynes and McKay have come up with a thorough list of the possible arguments in support of and opposed to the idea of implementing a plan to terraform Mars. These are listed below, either quoted or paraphrased directly from their work.

Arguments in favor of terraforming Mars and introducing an active biosphere include:

1. Even a thick, nonbreathable CO_2 atmosphere would simplify life for future astronauts or colonists; they could explore the planet equipped only with breathing apparatus rather than full environmental space suits.
2. Locally generated biomass would be an important source of energy, food, and other useful materials for astronauts or colonists.
3. Such an activity would provide a long-term project on which humans could focus, with a goal that was both useful and desirable for humans.
4. Such a project would seem to be an essential prerequisite for any future human colonization of Mars.
5. An active biosphere on Mars would provide a refuge for life on another planet in the solar system in the event of war or natural global catastrophe that might destroy all life on Earth.
6. Even a feasibility study of terraforming Mars, let alone its implementation, would generate important scientific and technological advances, stimulate new educational developments and economic activity, and foster international cooperation.
7. Much of the research involved would be highly relevant to addressing environmental problems on the Earth and understanding its biosphere.
8. If terraforming Mars and implanting an active biosphere is possible, then someday it will be initiated by somebody.
9. Planets with life on them have a greater intrinsic value than those without.
10. A commitment to a project such as this would provide a new frontier and a challenge for humans.
11. Solar system exploration and development is far less threatening than military development or an arms race, and would provide an outlet for international competition or technology development.

Arguments against terraforming Mars and introducing an active biosphere include:

1. The time scale for implementation is very long, perhaps longer than the lifetimes of the governmental institutions and world economic order needed to maintain the necessary commitment to such a project.
2. It is not clear that there are significant economic benefits, especially in the short term, that would be commensurate with the cost and effort involved.
3. Scarce human talent and economic resources would be diverted from other worthy projects such as addressing social and terrestrial environmental problems.
4. It is impossible to prove conclusively that Mars is totally devoid of life today; thus, the project should not be initiated for fear of extinguishing an indigenous martian biota.

5. It is more desirable to preserve Mars in its present state for scientific exploration, for aesthetic reasons, or for other future uses.
6. Something could go wrong in the course of the project that could damage the new martian biosphere beyond repair, leaving us worse off than before.
7. Humans have done such a bad job of managing the Earth's environment that it is presumptuous to imagine that they can be wise and successful planetary engineers on another world.
8. If terraforming were successful, Mars might become a tempting target for military and/or economic exploitation; this could generate more sociopolitical problems than we have at present.
9. Insurmountable political or legal roadblocks might exist.
10. The future evolution of a martian biosphere would be inherently unpredictable; contingencies exist that might be detrimental to humans or to Earth.
11. Such a project is, in some sense, sinful and would merit divine retribution.

The issues raised on both sides should be considered as discussion points rather than deciding factors. Rather than trying to reach a conclusion here, it is more appropriate that each individual consider these moral and ethical issues and reach their own conclusions. As it is not feasible to consider terraforming Mars at the present, however, there is sufficient time to debate and discuss the issues in detail.

Unintentional release of terrestrial organisms into the martian environment

An issue closely related to the intentional release of terrestrial biota into the martian environment is the unintentional release. This most likely would occur during spacecraft or human exploration of Mars, with terrestrial organisms that are inadvertently transported to Mars. This question of the biological contamination of Mars is extremely important, at least during the present period in which we are trying to determine if Mars has its own biosphere.

The 'forward-contamination' issue has been the subject of much discussion, going back to a 1967 international treaty that addresses it. That treaty, which was signed by both the United States and the (then) Soviet Union, included an agreement not to allow Mars to be contaminated with terrestrial biota during the period of biological exploration. Although it was initially felt that this period would last only 20 years, it is clear now that it will continue for the foreseeable future.

Initially, this agreement was implemented by requiring that a spacecraft sent to Mars have a probability of less than one part in a thousand of contaminating the planet with terrestrial biota. This goal could be accomplished by a variety of different approaches. For example, a spacecraft that would orbit the planet would need assurance that there would be an extremely low probability of hitting the planet during the orbit insertion maneuvers, in combination with a lifetime in orbit that would keep the spacecraft from crashing into the surface for at least several decades. For a spacecraft that would land on the surface, this goal would seem to require a high probability of a successful landing on Mars combined with a very low number of organisms living on the spacecraft that could contaminate the planet.

The two *Viking* spacecraft that landed on Mars in 1976, for example, underwent an alcohol wipe-down of the external surfaces of the spacecraft and of the individual components of the spacecraft, in order to remove any biological organisms. It was assembled in specially outfitted cleanrooms, by technicians in surgical garments, in order to minimize the contamination. In addition, the entire spacecraft was heated to try to reduce the number of living bacteria in its interior. Although this heating was not sufficient to completely sterilize the interior of the spacecraft, it was able to reduce the 'bioload' by a factor of approximately 10 000. The combination of these two approaches was deemed satisfactory, especially in the light of the successful landing and operations of the spacecraft. There was no indication that any of the experiments were in any way contaminated or compromised by terrestrial biota, or that any Earth organisms survived in the martian environment.

It is clear now, in fact, that the martian environment is not conducive to the growth and reproduction of terrestrial organisms (at least at the two *Viking* landing sites). The ultraviolet radiation incident on the surface is intense enough to kill most bacteria, and the oxidants that are thought to be present within the soil would destroy any organic molecules.

The modern approach to planetary protection is to require a low 'probability of growth' of terrestrial organisms. Thus, for a landing in most martian environments, it would be sufficient to reduce the bioload of the spacecraft by building it in a cleanroom and wiping down the outside of the spacecraft. However, for a lander that was intended to look for active biology, perhaps in an environment such as a hot spring where life could exist and where terrestrial organisms might survive, this is insufficient. There, more severe cleanliness requirements must be met. Of course, this is appropriate, since the purpose of life-detection or life-related experiments is to sample the martian environment rather than to sample terrestrial life that had been transported to Mars by spacecraft.

[Figure 10.3]
The 'face on Mars', just to the right of the upper middle part of the image. Nearby are features that are thought by some to look like a fortress and pyramids. The 'doughnut' to the left of the face is an instrument artifact. (NASA photo.)

The face on Mars

Finally in our discussion of possible life on Mars, it is appropriate to address the questions that have been raised surrounding the possible existence, at some time in the past, of a martian civilization. The evidence that has been cited for such a civilization is the presence of possible artifacts on the martian surface and, in particular, the so-called 'face on Mars'.

The feature referred to as the face on Mars is a small mountain located near the boundary between the heavily cratered southern-hemisphere highlands and the younger, volcanic, northern-hemisphere lowlands (Figure 10.3). It is about a kilometer across, and has the appearance of a human face, wearing a helmet (or perhaps having 'Prince Valiant' hair) and staring straight up into space (Figure 10.4). It is in a region that dates back geologically about two billion years, based on the number of impact craters in the surrounding area. There are numerous other isolated mesas and mountains in the vicinity, which are consistent with this region being at the boundary between the highlands and lowlands.

Certainly, there is no doubt that it looks like a face. In fact, it was first discovered and described as a face by Toby Owen, a *Viking* mission scientist (now at the University of Hawaii) who first discovered it while looking at images of the surface in 1976. However, it is unlikely that it owes its existence to intelligent creatures rather than to geologic processes. In fact, this is an excellent example of the proper application of the scientific method.

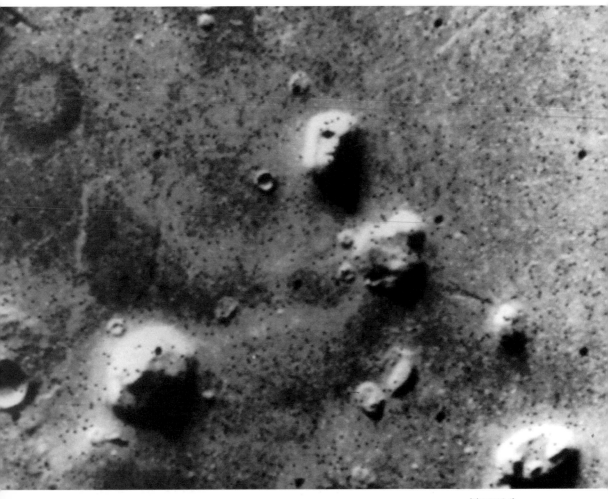

[Figure 10.4]
Close-up view of the 'face on Mars'.
(NASA photo.)

In trying to understand the origin of a feature such as this, one should look at the most likely and most plausible processes first; only as these can be eliminated should one consider less likely or plausible processes.

In this case, it is quite plausible that geological processes are responsible for producing the face. There are numerous remnants of the highlands boundary in the same region that are of similar size and shape, with only the specific details of the features differing. The fact that one bears a striking resemblance to a face must be treated as coincidence – there is a general tendency for humans to see familiar items in the world around them, whether it be faces in the clouds or the Moon, mountains formed in the image of animals or people, or images of gods in the constellations.

One cannot demonstrate that the face on Mars was not carved by intelligent beings some billions of years ago, just as one cannot rule out the possibility that the Earth is being visited by aliens today in the guise of UFOs. However, in the absence of concrete physical evidence that specifically and explicitly points to this as an explanation, the alternative (and very plausible) explanation of ongoing natural geological processes has to be considered to be the best interpretation.

171

Concluding comments

In this chapter, we have discussed a number of issues that are related to the question of life occurring on Mars. The issues that were raised – terraforming Mars, planetary protection, and the face on Mars – are outside of the realm usually covered in discussing exobiology. These topics have such widespread interest, however, and are very visible in today's culture. In fact, to the extent that these issues may be the only exposure that some people have to the possibilities of extraterrestrial life, they even could have been the starting point for the entire discussion. They certainly fit within the scope of the topics considered here.

However, just as certainly, the issues regarding terraforming Mars and the Face on Mars seem to be more within the realm of science fiction than science fact. The simplest and best explanation for the Face is that it is an erosional remnant created by natural geological processes, rather than something carved by ancient travelers. And, all of the available information suggests that terraforming Mars would be incredibly difficult, if not impossible. The desire to find evidence for intelligent life, or to change worlds to accommodate human life, really falls outside of the usual arena for science.

172   The planet Venus is not a likely place to find life today. Although it is just a little closer to the Sun than the Earth, and is shrouded in clouds that can reflect a large fraction of the sunlight, the surface temperature is about 750 K (477 °C, or 891 °F). This is too warm for liquid water to be stable at the surface – it would very quickly evaporate into the atmosphere – and the atmosphere itself is remarkably dry. At first glance, the surprisingly harsh conditions on Venus make a remarkable contrast with the Earth. Venus is about the same size and mass as the Earth and, with the exception of water, it seems to have a total inventory of volatiles such as carbon dioxide and nitrogen that is roughly comparable to that on the Earth. What is it that makes Venus so different from the Earth? Has Venus always had such a harsh environment? Might the planet have had a more-clement climate and harbored life at some time earlier in its history?

Despite its overall similarity to the Earth, Venus stands apart from it in two fundamental ways. First, Venus has an atmosphere that is almost 100 times thicker than the Earth's and is composed primarily of CO_2. This thick atmosphere induces a high surface temperature due to the greenhouse effect. Second, although it presumably has interior heat sources comparable to the Earth's (from the decay of radioactive elements, possible freezing out of an inner core, and residual heat from accretion, as discussed in Chapter 3), it does not have global plate tectonics with a style similar to the Earth's. And, it shows evidence for occasional catastrophic volcanic resurfacing rather than the steadier resurfacing that occurs on Earth. These differences make Venus a fundamentally different place from the Earth. Also, they seem to point out the importance of location in the solar system as a factor in controlling the behavior of a planet, in addition to a planet's intrinsic properties. We will discuss these differences from the Earth as well as the ways in which they interact to make Venus an inhospitable planet. We will see that these questions of planetary climate and history are fundamentally important when we look at planets around other stars and whether they could be habitable.

The Venus atmosphere and climate

The thick clouds in the Venusian atmosphere successfully shielded the planet's surface from view until about 30 years ago. Since the first spacecraft flyby of Venus in 1962, there have been a number of flybys, orbiters, and landers; in addition, imaging radar observations (both from Earth and

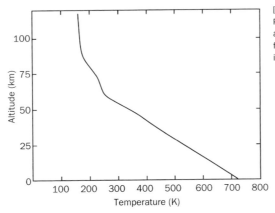

[Figure 11.1]
Profile of temperature through the
atmosphere of Venus, measured
from the *Pioneer Venus* entry probe
in 1978. (After Seiff, 1983.)

from spacecraft) have allowed the surface to be mapped despite the clouds. Although there is still much that we do not know about Venus, we now are able to put it into its proper place in understanding the solar system.

Venus is just a little bit smaller in size than the Earth and just a little bit less massive. Because it is close in size and mass, it is often considered to be the Earth's twin. Because of the much higher temperatures, however, it has been referred to as Earth's 'evil twin' . The atmospheric pressure at the surface is about 90 times the Earth's, and the atmosphere consists primarily of carbon dioxide. The temperature at the surface is more than twice as hot as the Earth's surface temperature (measured in absolute units), being about 750 K (Figure 11.1).

At an altitude of about 60 km above the surface, there are several layers of thick clouds that span over 20 km from top to bottom. Coincidentally, the base of these clouds is located at an altitude where the pressure and temperature of the atmosphere are similar to those at the surface of the Earth. The clouds consist primarily of sulfuric acid droplets. One of the important gaseous minor constituents in the Venus atmosphere is sulfur dioxide (SO_2), presumably derived from volcanic eruptions at the surface, and chemical reactions in the atmosphere are able to turn some of this gas into sulfuric acid. The total mass of cloud material is not very large, only perhaps 0.012 grams/cm^2; this is enough to form a layer of liquid about 0.1 mm thick, if it were all to condense onto the surface. This is enough, however, for the clouds to keep almost all of the sunlight that hits them from reaching the surface.

The clouds show striking patterns when viewed in the ultraviolet

(Figure 11.2). At these wavelengths, small amounts of sulfur-bearing particles above the clouds can absorb some of the reflected sunlight and
produce bright and dark patterns. These patterns are in the shape of a side-
ways V or Y, and can be used to track the winds at these altitudes. The winds
move at an average speed of 100 m/s, and the atmosphere at this altitude
will travel all the way around the planet in only four days. This is remarkable, given that the planet itself rotates on its axis only once every 243 days.
In fact, the planet rotates in a retrograde direction, the opposite direction
from the Earth's rotation; it is the only planet in the inner solar system that
does this. The length of a Venus year is only 224 days, and this combines
with the planet's own rotation to give a period of 117 days between one
sunrise and the next.

The high surface temperatures result from the greenhouse effect due to
the thick atmosphere. The CO_2 in the atmosphere is more efficient at
letting sunlight pass through it, despite the presence of the clouds, than it
is at letting out thermal-infrared heat energy. This causes heat to build up
in the lower atmosphere and increases the temperature. The temperature
rises to high enough values that the heat lost through radiation to space
matches the amount of incoming sunlight. The presence of a thick greenhouse atmosphere was inferred in the early 1960s, when the first radio
measurements of Venus showed high surface temperatures. However, only
recently have the details of the greenhouse effect been worked out. It turns
out that the atmospheric CO_2 alone cannot produce such a thick greenhouse. CO_2 does not absorb energy at all of the infrared wavelengths, so
some of the heat would be able to escape to space. If CO_2 were the only
greenhouse gas present in the Venus atmosphere, the surface temperature
would be about 100 K cooler than it is.

Other gases, however, can absorb energy at the wavelengths where CO_2
does not absorb. Water vapor, if it were present in sufficient amounts,
would be able to absorb the rest of the heat. However, water would need to
constitute about 0.5% of the atmosphere to be able to do the job, and it is
present only at an abundance of about 30–100 ppm (parts per million) or
0.003–0.01%. Atmospheric SO_2, CO, and HCl gases, in the relatively small
abundances in which they are present in the atmosphere, are able to
provide the remainder of the greenhouse warming.

It is interesting to note that the global Venus inventories of nitrogen
and carbon dioxide are similar to the Earth's. The Venus atmosphere consists of 96.5% CO_2 and 3.5% N_2, with minor amounts of trace species such
as argon, carbon monoxide, and so on. Thus, the total amount of CO_2
present in the atmosphere produces a pressure of about 87 bars. On Earth,

[Figure 11.2]
Spacecraft image of Venus,
taken using ultraviolet light. The
features that are seen occur at
the level of the tops of the
clouds, about 70 km above the
surface. (NASA photo.)

the total inventory of CO_2 would include that which is present in the atmosphere and dissolved in the oceans, as well as CO_2 present at the surface in the form of carbonate minerals. The carbonate minerals, such as calcite ($CaCO_3$) in limestone, were formed by precipitation of CO_2 gas that had been dissolved in the oceans. On the Earth, of course, all of these reservoirs are connected through a carbon cycle – limestone and other carbon-rich sediments are subducted at plate boundaries, the CO_2 is released back into the atmosphere by volcanism at subduction-zones, it dissolves from the atmosphere into the oceans, and is redeposited as seafloor sediments. The total amount of carbon participating in the Earth's carbon cycle is the equivalent of about 60 bars of CO_2, very close to the amount of CO_2 on Venus. Similarly, the Venus atmosphere contains about 3 bars of N_2, and the Earth's total inventory of N_2, including nitrogen in the atmosphere, biosphere, and sediments, is equivalent to about 3.8 bars.

In contrast, however, the Venus water abundance is substantially less than the terrestrial inventory. All of the Venusian water is in the atmosphere, since the surface is too hot to allow either liquid water or hydrated minerals to exist. At a level of about 100 ppm, there is only about 10 mbar of water vapor. In contrast, the Earth's oceans are an average of 3 km thick, which is equivalent to a global layer of water about 2 km thick. If all of the Earth's water were put into the atmosphere as vapor, it would be the equivalent of roughly 100 bars of atmospheric pressure. It is hard to imagine a process either by which the two planets would have formed originally with such different amounts of water, or by which Earth would be able to outgas the water to the surface but Venus would not. Therefore, it is thought that Venus must initially have had an amount of water comparable to the Earth, but that it lost its water over time to space. One process by which the water could have been lost would have been the photodissociation of water by sunlight, whereby a water molecule is broken into its component atoms, followed by the escape of the hydrogen to space. The oxygen released by this process presumably would have oxidized surface minerals.

What might be the cause for such different climates on Earth and Venus? One important mechanism may be what has been referred to as a 'runaway greenhouse'. It is clear from numerical calculations that the gases present in the Venus atmosphere are able to retain enough energy to keep the temperature at its current high value. What would happen if an Earth-like planet, with its oceans and its 1-bar oxygen–nitrogen atmosphere, were placed at Venus' distance from the Sun? The atmosphere

[Figure 11.3]
Model of the temperature behavior of a greenhouse atmosphere for an Earth-like planet. The greenhouse model includes the radiative effects of CO_2 gas and of water vapor that evaporates from a global ocean. The effective solar flux is a measure of the efficacy of the greenhouse in trapping energy, and can be thought of as a measure of the total atmospheric pressure; the present Venus greenhouse corresponds to a value of about 1.6, but the actual Venus atmospheric temperatures are not as high as indicated due to the absence of large quantities of water vapor. (After Kasting *et al.*, 1993.)

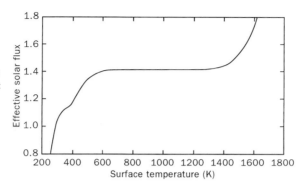

would begin to warm up, because of the higher input of solar energy; the temperature would increase by perhaps 10%, or about 30 K. At the higher temperature, the atmosphere would be capable of holding more water vapor, and some water from the oceans would evaporate. Water vapor is a greenhouse gas, and absorbs heat energy at wavelengths where CO_2 does not; the increase in atmospheric water vapor, therefore, would cause the atmospheric temperature to increase a little bit more. This additional temperature increase would cause the atmosphere to be able to hold additional water vapor, which would evaporate from the oceans. This would cause an additional heating, and this cycle would repeat itself in a runaway effect. The system would not stabilize until all of the water from an ocean were in the atmosphere and the temperature was at very high values (Figure 11.3).

Why doesn't the Earth show this same runaway greenhouse? At the lower terrestrial temperatures, the amount of water vapor that the atmosphere can hold is much less than if it were even a little closer to the Sun. As a result, the additional increase in the greenhouse effect from the water vapor is smaller, and the feedback is less (Figure 11.3). The water vapor provides a small additional heating, but not enough to trigger a runaway effect. As a result, the Earth's atmosphere is stable at relatively low water abundances and temperatures. The Earth would need to receive about 40% more sunlight than it currently does in order to trigger a runaway greenhouse. This extra sunlight would hit the Earth if, for example, it were closer to the Sun, at a distance of about 0.85 A.U. For comparison, Venus is at a distance of 0.72 A.U. from the Sun, well inside the boundary for a transition to a runaway greenhouse. The same result would also occur if the Sun's energy output were greater than it is today; the Sun's output is increasing with time, and we can expect the Earth to reach the critical stage of initiating a runaway greenhouse in perhaps only two billion years. As we will see in Chapter 16, the possible occurrence of a runaway greenhouse places important constraints on where a planet in a solar system around another star can be in order to be habitable.

Once the water on Venus has evaporated into the atmosphere, it can be lost to space (as described earlier). And, there is observational evidence that substantial amounts of water have escaped to space. Normal hydrogen (H) has only a proton in the nucleus, while the naturally occurring but less abundant heavier isotope of hydrogen, deuterium (D), has both a proton and a neutron. It is twice as heavy as hydrogen, and will not be able to escape to space quite as easily. As hydrogen escapes to space, H escapes more efficiently than D, and the hydrogen gas that is left behind becomes enriched in deuterium. The measured ratio of deuterium to hydrogen, D/H, in the Venus atmosphere is about 2.4×10^{-2}, some 150 times the terrestrial value of 1.5×10^{-4}. This enrichment in D/H argues that a substantial amount of water must have been lost to space over time. The 30 ppm of water vapor currently in the Venus atmosphere therefore must be a residual amount left over after a larger amount has been lost. In fact, it would suggest that we are lucky to see any water in the Venus atmosphere because, at the current hydrogen escape rates, all of the water present today in the Venus atmosphere will be lost in only about 200 million years.

We can use the 150-fold enrichment in D/H to estimate the initial amount of water that must have existed on Venus. The best estimate currently suggests that Venus must have had at least enough water to create a global layer about 3 m thick, or about 0.15% of the Earth's water inventory. This estimate has a large uncertainty, however, because water can be resupplied to the Venus atmosphere by the impact of water-ice-rich comets; every comet impact will provide water that has an unaltered D/H ratio, and the water supplied will partially 'reset' the planet's D/H ratio to nearly the value in the comet ice. The total amount of water in the Venus atmosphere today is about 3.8×10^{18} g, equivalent to a hypothetical global layer of liquid about 1 cm thick. This amount of water could be supplied by the impact of about 10–20 comets that are the size of Halley's comet. Such impacts occur regularly in the solar system, as recalled from our discussion in Chapter 2 of the impact rate and its effects on the Earth. Impacts of objects this size would be expected to occur about every 10–100 million years. This means that the Venusian water vapor should have been entirely resupplied during the history of the planet. The present measured value of D/H then would have to result from a balance between loss to space (which increases the atmospheric D/H) and resupply from comets (which resets it). This means that we cannot easily determine exactly how much water has been lost through time; the above estimate has to be considered as a lower limit, and the observations could easily accommodate an initial water inventory of 1% of the Earth's oceans or greater.

The Venus surface and interior

The nature of the surface and interior of Venus has only begun to be understood within the last decade. Our understanding is based largely on the images of the surface obtained from radar observations, since the radar waves are not affected by the pervasive cloud cover and can see through to the surface. Although substantial radar imaging has been done from the Earth, using radar facilities such as those at Arecibo in Puerto Rico and Goldstone in California, the highest quality, the best spatial resolution, and the most global coverage comes from radar measurements that were made from the *Magellan* spacecraft. *Magellan* went into orbit around Venus in 1990, and returned data for several years before it entered the Venus atmosphere and crashed onto the surface. Analysis of the data is still ongoing, and any discussion of the results has to be considered as preliminary.

In order to understand the Venus radar images, we first need to understand how imaging with radar differs from imaging using reflected sunlight. In the *Magellan* radar experiment, radio waves were transmitted by the spacecraft toward the Venus surface, and they hit the surface and were reflected off of it. Some of the radar energy bounced back in the direction of the spacecraft, where its properties could be measured. The way in which radar signals bounce off of the surface depends upon the roughness of the surface. A smooth surface will reflect radar much like a mirror; and, since the radar beam was aimed at the surface at an oblique angle, the reflected beam would be scattered away from the spacecraft – the surface would appear dark in the resulting radar image. A rough surface, on the other hand, will scatter the radar energy into all directions, and some of the energy will be scattered back toward the spacecraft – the surface would appear bright in the radar image; this is similar to the way in which a movie screen will appear bright when a light shines on it, regardless of the direction from which it is being viewed.

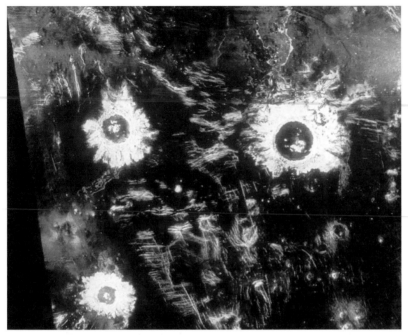

[Figure 11.4]
Three impact craters on the surface of Venus, observed with radar from the *Magellan* spacecraft. Each crater is about 50 km across at the rim, and is surrounded by bright debris ejected by the impact. (NASA photo.)

When we speak of a surface being either 'smooth' or 'rough', we are referring to its properties at the physical scale of the radar wavelength, in this case 12.5 cm. Thus, an asphalt street would be smooth at this scale, since all of the physical roughness is at a scale of around 1/2 cm or less. A pile of bricks, on the other hand, would be rough at this scale, since the bricks themselves are all about the same size as the wavelength of the radar. Geologically, we would expect to find both smooth and rough areas on the surface. For example, some lava flows are smooth, such as those which are similar to terrestrial pahoehoe lava flows; these have a smooth, glass-like surface, as they were formed from a very fluid lava. And some lava flows would be rough, like terrestrial aa flows; these have a very rough surface, which is difficult to walk over.

The radar images clearly show impact craters (Figure 11.4). They range in diameter from a few kilometers up to about 280 km. They can be recognized as impact craters by their circular shape, the size and appearance of the debris that surrounds the circular rim, and the occurrence in the larger craters of rebound peaks in the middle of the crater. Although there are some differences, these features are very similar to what is found in impact craters on the Earth, the Moon, and Mars. Craters smaller than a

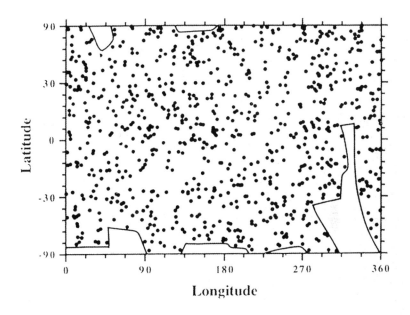

[Figure 11.5]
The distribution of impact craters over the surface of Venus. Each dot represents a single impact crater, and the outlined areas are locations where no data exist. (From Schaber *et al.*, 1992.)

few kilometers in diameter generally are not found, because the objects that would create them are too small to get through the thick Venus atmosphere.

There are about 950 impact craters on Venus (Figure 11.5). This is about four times as many impact craters as are found on the Earth. Geologic processes on the Earth are able to destroy the craters more rapidly than on Venus. On Earth, the ocean floor, which covers about 3/4 of the planet's surface, is destroyed and created anew every 150 million years or so. On the continents, the primary process for obliterating impact craters is erosion by water; this can remove the topography associated with a crater (the raised rim and depressed bowl center) in anywhere from 1 to 50 million years. On Venus, there is no rainfall and the atmospheric winds at the surface (only a couple of meters per second, much less than the high-altitude 100 m/s winds) are not able to erode away very much of the surface. There appears to be no global plate tectonics of the style seen on Earth, as will be discussed in detail below. This leaves burial by volcanic lava flows and distortion by tectonic movement as the major mechanisms by which craters are destroyed.

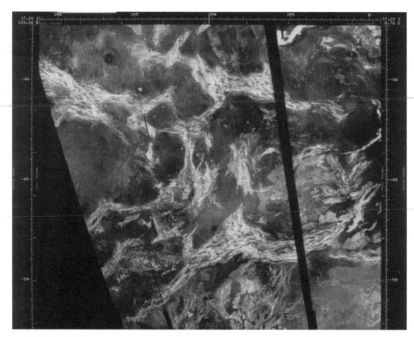

[Figure 11.6]
Regions on Venus that have
been deformed by tectonic
motions. The bright regions
show where the surface has
been deformed. The dark
regions between them are
volcanic. The image is
about 1800 km across.
(NASA photo.)

Based on our estimates of the rate at which objects hit the surface, the impact craters on Venus would have been created in about 500 million years, and the average age of the surface therefore is 500 million years. Importantly, the geographical distribution of the craters over the surface appears to be random or very close to random (Figure 11.5). As a result, there are no large areas that are much older than or much younger than this. There can be older or younger areas, but they cannot be much larger than the average spacing between impact craters. This spacing is about 600 km, equivalent to an area about 5 degrees in latitude in diameter or about the distance between Los Angeles and San Francisco.

There does not appear to be a global network of spreading centers and subduction zones that would suggest that terrestrial-style plate tectonics is occurring on Venus. If Venus did have the same types of subduction zones and spreading centers that are seen on Earth, they would show up clearly both in the radar images and in global topography measurements. There are large-scale tectonic features, however, that do show that some significant deformation has occurred at the surface (Figure 11.6). These show continent-sized zones where the surface has been deformed by movement, presumably driven by the motions of convection cells in the

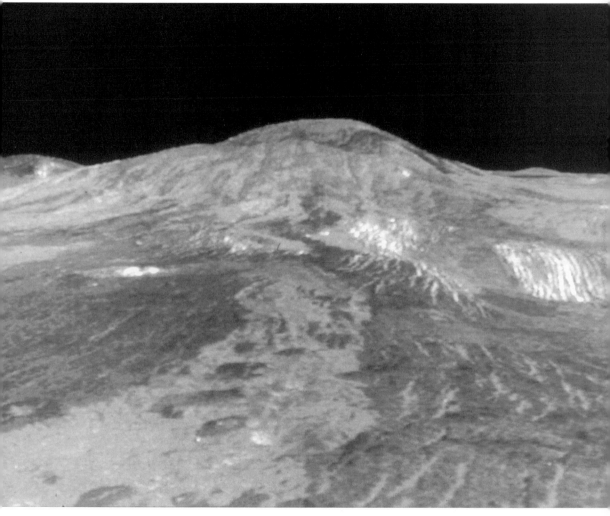

[Figure 11.7]
Magellan radar image of the volcano Sif Mons on Venus. The view is a 'perspective' view, showing how the volcano would look when seen obliquely (with a 10× exaggeration in vertical relief). Notice the lava flows leading away from the mountain. (NASA photo.)

upper mantle. These do not appear to have spreading centers or subduction zones associated with them, however. Much of the surface appears to be volcanic in origin, with abundant lava flows, flood eruptions, and central-vent volcanoes (Figure 11.7).

There are features that have characteristics somewhat similar to those associated with plate tectonics, but on a much smaller scale. Features called 'coronae' (or, in singular form, corona) are approximately round features that are up to several hundred kilometers across (Figure 11.8). They have what appear to be faults or zones of tectonic deformation around the

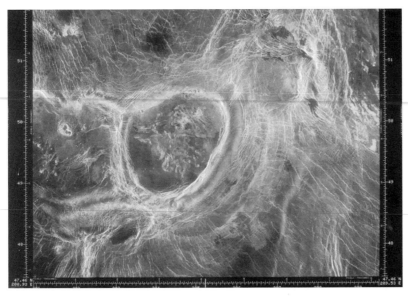

[Figure 11.8]
A Venusian 'corona', a nearly circular region of deformation that has been filled in by volcanic eruptions. The corona is about 150 km across. The deformation at the edge combined with the upwelling in the center that produces the volcanism may indicate that these features result from small-scale convective overturning. (NASA photo.)

perimeter, and the interior regions are filled with volcanic lavas. One possible mechanism for their formation involves the presence of a small-scale circulation cell, with mantle upwelling in the middle and downwelling at the edge. The upwelling material brings magma with it that erupts to the surface to form the volcanic interior; the downwelling regions may be similar to subduction zones. The outer edges of the corona even have topography that mimics that which is seen at terrestrial subduction zones, with a trench and a rise associated with it. The trench-like features associated with the coronae are the same size as the smallest subduction zones on the Earth, such as the isolated subduction zone in the south Atlantic ocean (Figure 11.9). The integrated length of the possible trench-like features in the coronae (and in isolated chasms that show similar properties) is about 10 000 km. This is less than half of the integrated length of subduction zones on the Earth. Thus, if subduction is occurring on Venus, it is much less widespread than on Earth.

Venus also does not have the clear distinction between continents and ocean basins that the Earth has. On Earth, the distinction is both topographical and compositional, with the continents consisting of recycled ocean-floor material and standing several kilometers higher than the ocean floor (see Chapter 2). While Venus does have large variations in topography, with the range in altitude from the lowest to the highest sur-

[Figure 11.9]
Comparison of topography at a suspected subduction trench on Venus with topography at a trench on Earth. In each case, the dashed line shows the topographic relief that is expected at a subduction zone. (After Schubert and Sandwell, 1995.)

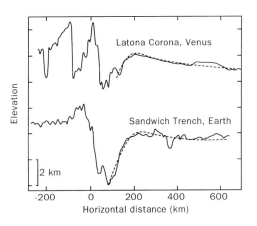

[Figure 11.10]
Histogram of the distribution of topography on the Earth and on Venus. The elevation scale has tick marks at 2-km intervals, and the curves show the fraction of the surface located at each elevation. The curves for Venus and Earth have been matched vertically to their mean values, indicated by the arrowhead. Earth has two distinct modes, corresponding to the continents and the oceans. Venus has only a single mode, indicating that it is not divided into ocean-floor and continental material in the same way. (After McGill et al., 1983.)

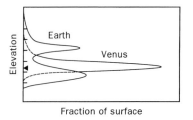

faces of about 20 km being comparable to the range on Earth, most of the large regions of higher topography do not have the sharp edge that marks terrestrial continents (Figure 11.10). Also, the regions of high and low topography do not show the same relationship to tectonic features as they do on Earth.

We see that Venus as a planet is substantially different from the Earth. It is about the same size and mass as the Earth, and we expect it to have about the same composition as the Earth. We imagine that it most likely would have differentiated into a core, mantle, and crust that are similar to those on the Earth, but we do not have any measurements that tell us directly that this has happened. We expect that it has similar heat sources in its interior and that, as a result, the mantle should be convecting. However, we do not see much evidence at the surface for a convecting mantle – no global-scale plate tectonics, and a limited number of what may be local-scale convection and subduction zones. Furthermore, the evidence from the number and distribution of impact craters on the surface suggests that the planet is not being continuously resurfaced on a large scale.

Altogether, the observations suggest that the Venus surface was resurfaced about 500 million years ago and has been relatively stagnant since then. This implies that Venus might have periods of minimal activity, with episodes of catastrophic volcanic resurfacing occurring at some interval. What might cause this difference from the relatively steady resurfacing of the Earth? We know two ways in which Venus differs from the Earth that might affect the plate tectonics. First, the surface temperature is substantially hotter on Venus than on Earth. This would lead to a warmer interior. And, because physical properties such as the viscosity of rock in the mantle depend on the temperature, we would expect that the Venusian mantle will be less viscous than the terrestrial mantle. Second, the absence of water vapor in the Venus atmosphere means that the mantle of Venus will be drier as well – on Earth, subduction of water-rich sediments continually replenishes the water in the mantle. The lack of mantle water on Venus also might affect the viscosity, making the mantle more viscous than the Earth's mantle. The combination of these two effects might introduce a 'non-linearity' in the response of the Venus interior to heating. The Venus interior might be able to heat up substantially without any convective overturning of the mantle occurring. Then, when the interior temperature reaches some high value, the mantle would suddenly begin to convect, letting out all of the heat and allowing the interior to cool. Global-scale surface volcanism that would completely resurface the planet would be associated with this convection. Once the heat is released and the interior cools off, the mantle would stop convecting again and the surface would become stagnant. Thus, there would be a periodic catastrophic resurfacing of the planet.

Chemical reactions and possible instability of the Venus atmosphere

Now that we see that the surface can undergo periodic changes associated with the interior, let's examine some of the ways in which the surface and the atmosphere can interact, and the effects on the long-term stability of the atmosphere. This will lead us toward a discussion of possible biological activity that might be connected to the climate history.

The gases in the atmosphere are in contact with the rocks on the surface, and they can interact with them chemically. Conceptually, this is similar to the way in which iron at the Earth's surface will interact with

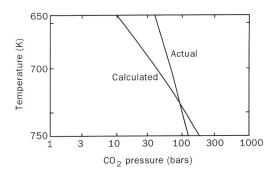

[Figure 11.11]
Temperature versus pressure as actually measured in the Venus atmosphere and as expected based on mineral equilibrium chemistry. The similarity of the values at the bottom of the atmosphere (pressure of 92 bars) suggests the possibility that the surface pressure is controlled by the mineral chemistry. (After Fegley and Treiman, 1992.)

water and oxygen in the Earth's atmosphere and form rust. The Venus chemical reactions, though, are somewhat different. For example, one reaction which might be important is the formation of the mineral calcite from CO_2 in the atmosphere:

$$CO_2 + CaSiO_3 \leftrightarrow CaCO_3 + SiO_2. \tag{1}$$

Here, atmospheric CO_2 can react with the mineral wollastonite ($CaSiO_3$) to form calcite ($CaCO_3$) and quartz (SiO_2). The double arrow indicates that this reaction can proceed in either direction, depending upon the pressure and temperature at the Venus surface. What makes this reaction particularly interesting is that, at the Venus surface temperature of 750 K, the reactions in each direction will have the same rate when the CO_2 pressure is about 90 bars; this is almost identical to the measured pressure (Figure 11.11). At this pressure, then, the *net* reaction rate, which is the difference between the rates of the reactions in the two directions, would be zero and the system would be in equilibrium. Another way of saying this is that, at the Venus surface temperature, the chemical reactions will either use up or release CO_2 in order to drive the pressure toward a value of 90 bars. It has been suggested that this chemical reaction, in fact, might be responsible for setting the Venus atmospheric pressure at its present value.

However, although the current pressure is close to the equilibrium value for this reaction, it is an unstable equilibrium. The reason is that there is also a feedback between the CO_2 pressure and the magnitude of the greenhouse warming. If the temperature remained fixed, a small change in CO_2 pressure in the atmosphere (for instance, by a sudden volcanic eruption) would change the forward and backward reaction rates and cause the pressure to return to its equilibrium value. With the temperature feedback, however, a small change in the CO_2 pressure would cause a change in the surface temperature that would then trigger an even larger change in the pressure. This creates an unstable situation in which the pressure is driven *away* from the value of 90 bars. In this type of an unstable chemical system, there are only two stable values of atmospheric pressure. At the high-pressure end, the system would not stabilize until all of the CO_2 is in the atmosphere and any surface reservoir of calcite is completely used up. At the low-pressure end, these reactions will stop at a pressure of about 3 bars (down from the current 90 bars) and a temperature of about 350 K; in this case, the rest of the CO_2 would be in the form of surface deposits of calcite.

What makes this particularly interesting is that a change in the amount of the greenhouse gas SO_2 could also trigger this instability. SO_2 is put into the atmosphere by volcanic eruptions, and can be removed by a chemical reaction with calcite,

$$SO_2 + CaCO_3 \rightarrow CaSO_4 + CO. \qquad (2)$$

(The mineral $CaSO_4$ is called anhydrite.) If a change in the volcanic eruption rate occurs, the change in the amount of atmospheric SO_2 would change the temperature of the atmosphere by a small amount. This would be enough to trigger a jump to an unstable atmosphere, and the atmosphere would evolve quickly to one of the two stable points. If the SO_2 abundance increased, it would drive the CO_2 to the high-pressure stability point, with all of the CO_2 in the atmosphere. If the SO_2 abundance decreased, it would drive the CO_2 to the low-pressure stability point, with most of the CO_2 being present as calcite on the surface.

For the CO_2 to be removed from the atmosphere, however, there must be sufficient wollastonite ($CaSiO_3$) with which it could interact chemically. Otherwise, the reaction would stop when all of the wollastonite had been used up. It is possible that the Venus atmosphere currently is in one of these extreme states, either with all of the CO_2 in the atmosphere or with some of the CO_2 as calcite in the surface but with no more wollastonite available for chemical reactions.

Or, we even could be seeing Venus in a transient state, with the atmosphere currently in the process of chemically reacting with the surface and forming calcite. If this is the case, then we would expect the atmospheric CO_2 to become depleted, and surface calcite to form, on a timescale of about 10–20 million years. Venus might undergo periodic oscillations of the atmosphere: CO_2 could form calcite on the surface; when the interior begins to convect and the surface undergoes catastrophic resurfacing, the calcite would be heated and would force all of the CO_2 back into the atmosphere. After the resurfacing event, the CO_2 might react again to form calcite on the surface. The formation of surface calcite could even *trigger* resurfacing by cooling the atmosphere and creating a large temperature difference between the interior and the surface.

Terraforming Venus

There has been the suggestion that we could terraform the Venus atmosphere, and create more clement temperatures. There are several possible mechanisms by which this could be done. For instance, we could move asteroids into orbits such that they would collide with Venus, with each asteroid impact ejecting some atmosphere from the planet; however, there are not enough large asteroids in the solar system to accomplish this, and it would require the devastation of a large fraction of the Venus surface by the force of the impacts. A simple alternative mechanism of reducing the temperature would be to use a sunshade to reduce the solar energy incident on the Venus surface, or to add dust to the upper atmosphere that would absorb or reflect sunlight before it reached the surface.

An interesting means of cooling off the surface of Venus might be with microbes that could turn the CO_2 into carbohydrates. There are bacteria that can do this on the Earth, and with only a small amount of genetic engineering they might be able to do so on Venus. The problem with this approach, however, is that carbohydrates or fully reduced carbon (elemental C) would react chemically with the oxygen that was produced and would return the CO_2 back to the atmosphere.

Of course, any of these mechanisms might change the amount of CO_2 in the atmosphere by just enough, and for just long enough, to trigger the atmospheric instability described earlier. This would cause the bulk of the atmospheric CO_2 to react to form calcite on the surface. If there is sufficient wollastonite with which the CO_2 could interact, then we would only need to wait for a few tens of millions of years to see a more Earth-like climate. Of course, this also could trigger the catastrophic overturning of the atmosphere, which would quickly undo our efforts!

Venus exobiology

Venus has an atmosphere and climate which is relevant to our discussions of exobiology, even though it is not conducive to the presence of life today.

It is clear, however, that Venus had more water at some time in the past. In addition, there are two possible scenarios in which the climate on Venus could have allowed the existence of life at some time in its past. First, the possible unstable nature of the present-day climate could allow Venus to have a more-clement atmosphere even at the present epoch. If most of the CO_2 were removed from the atmosphere and deposited onto the surface as carbonate minerals, the temperature could drop to as low as 350 K. Even with only a small amount of water available, life could exist at these temperatures. Second, with the energy output of the Sun having been perhaps 30% lower than its present value several billion years ago, Venus might have had a lower surface temperature then. Temperatures might have been much more like those on the modern Earth, and would have been low enough to allow the existence of life.

Concluding comments

When we look at the trio of terrestrial planets – Venus, Earth, and Mars – although we see only one planet that harbors abundant life at its surface today, the situation might have been very different 4 billion years ago. Then, the Venus atmosphere might not have undergone a transition to a runaway greenhouse yet, and the surface temperatures would have been conducive to the existence of life. And, the geological evidence for running water on the surface of Mars suggests that it also might have allowed life to exist at the surface then. Four billion years ago, life could have existed on all three planets! Over time, Venus heated up and became too hot for life, Mars cooled off and became too cold for life at the surface, and only the Earth continued to have a clement climate. Life even could have formed independently on all three planets. Or, given the possibility that living organisms could be exchanged between planets, carried by meteorites ejected by impacts (see Chapter 8), life could have originated on any one of these planets and then been transferred to the others!

Although it may have been possible for life to exist on Venus during a past, more clement epoch, the temperatures are too high today. Further, the temperatures are too high for any physical evidence of a biological past to remain at the surface. If life ever existed on Venus, or if the climate ever was more clement than with the present greenhouse atmosphere, we will never know about it. Thus, there are aspects of the early history of the solar system and the evolution of life that can never be determined.

However, we can use Venus and its climate to understand better the behavior of planets in general. It will serve as an example of how a planet responds to solar energy input in amounts greater than the amount the Earth receives. As we will see, this is an important boundary condition for where in a solar system a planet might have a habitable climate.

The inner solar system has been the focus of our attention up until now, with discussion centering on the present and past habitability of the terrestrial planets. Obviously, Earth is quite a habitable planet – indeed, it is the only one on which liquid water is both stable and abundant at the surface, and on which water vapor is present in the atmosphere and ice can exist on the surface. Venus is closer to the Sun than the Earth and the increased sunlight appears to have triggered a runaway atmospheric greenhouse effect; any early ocean that might have been present on the surface has long since evaporated into the atmosphere and the hydrogen atoms have been lost to space. The resulting thick, CO_2 atmosphere is too hot to allow either liquid water or life. Mars is farther from the Sun than the Earth and, therefore, is colder; it shows clear evidence that liquid water has flowed over the surface at various times in the past; since water is not present at the surface today, it is either marginally stable or only occasionally present at the surface. Remarkably, the three terrestrial planets span the entire range of habitability, from too close to the Sun to expect life to occur, to just right, to barely too far from the Sun. As we move farther into the solar system, we might expect that any planets or satellites would be much too cold to be of biological interest.

It is surprising, then, that we find a number of planetary satellites in the outer solar system that appear to have some of the environmental requirements for life. The origin or continued existence of life would seem to require the presence of liquid water, access to at least the major biogenic elements, and a source of energy to drive chemistry away from equilibrium. Although none of the planets or satellites in the outer solar system is known with any certainty to meet all of these requirements, several meet one or more of them. The planet Jupiter, the Jupiter satellites Io and Europa, and the Saturn satellite Titan meet at least one of them. In addition, the Jupiter and Titan atmospheres show abundant evidence that organic chemical processes have occurred that may be similar in some ways to the chemical processes that took place on Earth and led up to the origin of life there. We will start with a discussion of Titan.

Titan is large enough to be a planet in its own right, except that it is in orbit around another planet, Saturn. With a diameter of 5150 km, Titan is almost as large as Mercury. Since the 1940s, it has been known to have a thick atmosphere that obscures the surface from observations, and to contain methane (CH_4) as a gas in its atmosphere. The atmospheric pressure at the surface of Titan is half again greater than that on the Earth; indeed, Titan has one of the thickest atmospheres of the solid planets in our solar system, second only to Venus. (Although technically a satellite, Titan is for all practical purposes a planet in its own right and we will refer to it as such.) And, as we will see, the nature of the Titan surface and atmosphere, even though not well understood today, may be of significant biological interest within the planetary science community.

The nature of Titan

Having a thick atmosphere immediately makes Titan an intriguing planet (Figure 12.1). The combination of a thick atmosphere and dense clouds that obscure the surface suggested that Titan might have significant greenhouse warming. As on Venus (Chapter 11), the clouds and gas would let some of the sunlight through to the surface but would not let infrared energy escape as easily. If such a greenhouse existed, then it would be possible for the surface temperature to be well above the temperature expected for an airless satellite. At Saturn's distance from the Sun, 9.5 A.U., a typical satellite should have a surface temperature below 100 K. At such a low temperature, water would exist only in the form of ice rather than liquid, and there could be no biological activity. However, if a greenhouse atmosphere were present it could raise the temperature substantially, possibly enough to allow liquid water to exist.

Although Titan has been observed in great detail using Earthbased telescopes, the *Voyager* spacecraft flyby in 1982 provided the information that finally allowed us to begin to understand the planet. Voyager observations provided information on the size and mass of Titan and on the composition and structure of its atmosphere.

The radius of Titan has been measured using telescopic images, but the results include the thickness of the atmosphere since we cannot see through to the surface. The *Voyager* spacecraft radio occultation experiment, however, provided us with the radius to the surface itself. In an occultation, the spacecraft passes behind Titan, as seen from Earth, and the radio signal that is broadcast from the spacecraft toward the Earth is tracked as it is received on Earth. The radio signal passes through the atmosphere with only minor changes in its character, but it cannot pass through the solid planet. The time at which the signal disappears tells us the location of the actual edge of the solid part of the planet and, therefore, its radius. Further, when the spacecraft passed by Titan, its trajectory was deflected by a small amount due to Titan's gravitational pull; the amount of the deflection allows us to determine Titan's mass. From the mass and the radius, we can determine the average density of Titan – about 1.9 g/cm^3.

This single determination of the average density tells us much about

the structure and composition of Titan's interior. If Titan were a purely rocky planet, like Earth or the Moon, its density would reflect a rocky composition and would be more than 3 g/cm³. The Moon's average density is about 3.3 g/cm³, and is consistent with a pure rocky, silicate composition. The Earth's average density, about 5.2 g/cm³, reflects its dense, iron-rich core in addition to a rocky mantle and the compression of its interior that results from the weight of all of the overlying mass. At 1.9 g/cm³, Titan must have a substantial amount of something lighter in addition to the rocky material.

The most likely material that could decrease Titan's density is ice. Most ices have a density between about 1 and 1.5 g/cm³. Although we might expect to find (and in various places in the solar system do, in fact, find) ices made from water, carbon dioxide, nitrogen, methane, and ammonia, we expect water ice to be the most abundant of these. Hydrogen and oxygen are both very abundant in the universe and in the solar system, and they combine to form the water molecule under a wide variety of conditions. Water ice is stable at the cold temperatures that occur in the outer solar system. It is present on the surfaces of several of the larger Jupiter satellites, as seen in the absorption of certain characteristic wavelengths of reflected sunlight, as well as on Saturn's rings. And, ice is one of the dominant components of comets, comprising perhaps half of their mass. So it should not be too surprising to find it on Titan. Water ice has a density of around 1 g/cm³, although it will have a slightly greater density at the higher temperatures and pressures that occur deep within a planet. So, with a density of less than 2 g/cm³, Titan must contain a large amount of water ice. Close to half of its volume would need to be ice in order to produce the observed density (Figure 12.2). Substantial amounts of some of the other ices could exist on Titan as well, however, and we might expect to find both ammonia and methane ices there.

The atmosphere of Titan provides the other significant clues to Titan's history. Both the atmosphere's temperature structure and its composition were determined from the same radio occultation experiment that was used to determine the radius. The composition was determined in part by a measurement of the average weight of the molecules in the atmosphere, which can be derived from the degree that the atmosphere slows down the radio waves as they pass through. For Titan, the average molecular weight turns out to be 27 atomic mass units (a.m.u.), meaning that the gas molecules contain an average of 27 protons and/or neutrons. The common molecules that have a mass close to this are nitrogen (N_2) and carbon monoxide (CO), both with a mass of 28 a.m.u. Nitrogen is more likely to be abundant in the Titan atmosphere than is carbon monoxide, since most of

[Figure 12.1]
Titan, as seen from the *Voyager* spacecraft looking back after passing by it. Thick clouds obscure the view of the surface at visible wavelengths. (NASA photo.)

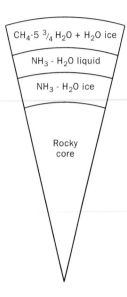

[Figure 12.2]
Possible interior structure of Titan, based on the observed density and the inferred surface composition. From the inside out is a rocky core, a mixture of ammonia and water ice, an ammonia–water liquid (deep subsurface ocean), and a mixture of methane clathrate hydrate and water ice. The methane clathrate hydrate is a solid structure composed of water molecules, with methane molecules filling 'holes' in the structure. Not shown is a thin surface layer (an ocean or isolated lakes) of liquid methane, ethane, and nitrogen. (After Lunine, 1993.)

the carbon in planetary atmospheres or ices is the more-oxidized carbon dioxide, and only trace amounts of CO have been detected in Titan's atmosphere. The presence of nitrogen has been inferred directly, using spectroscopic measurements. As a result, we expect that the bulk of the atmosphere is composed of nitrogen. The spectroscopic observations have indicated that other gases are present as well, including methane, ethane, and hydrogen, and they can comprise as much as perhaps 10% of the atmosphere.

The profile of temperature through the atmosphere was also determined from the radio occultation. The radio waves transmitted from the spacecraft were slowed down as they passed through the atmosphere, and the degree of slowing is determined by the density and the temperature of the atmosphere. The measurements suggest that the temperature at the surface is about 94 K, that it drops to nearly 70 K at around 40 km above the surface, and that it rises again at higher altitudes (Figure 12.3). This type of temperature structure is very similar to that of the Earth's atmosphere, where the temperature initially drops with increasing altitude above the surface but then rises again in the stratosphere. Although there is evidence for a greenhouse warming of the Titan surface of several degrees, the relatively low temperatures rule out the presence of a strong greenhouse effect. However, as will be seen, the chemical behavior of the atmosphere involves a number of organic molecules, even at these low temperatures;

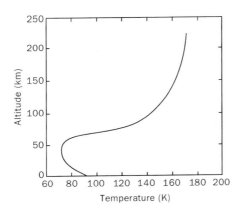

[Figure 12.3]
Atmospheric temperature profile
through the Titan atmosphere,
as measured from the *Voyager*
spacecraft. (After Lindal *et al.*,
1983.)

these make Titan an interesting natural laboratory for studying these types of processes.

Although the atmosphere of Titan is thought to be predominantly nitrogen, it also contains small amounts of methane (CH_4). The methane has been detected by using both reflected sunlight as measured by Earthbased telescopes and emitted energy as observed by the *Voyager* spacecraft. Can either the nitrogen or the methane be condensing in the atmosphere to produce the thick clouds that obscure the surface? The temperatures are not quite cold enough to allow nitrogen gas to condense, so the clouds cannot consist of nitrogen ice.

The issue of methane condensation is a bit more complicated. Methane is not stable in the atmosphere and would be expected to be lost from the atmosphere very rapidly – the molecules would be broken apart by ultraviolet radiation from the sun, even at Saturn's large distance from the Sun, and the hydrogen would escape to space. This process would be able to remove all of the present atmospheric methane in only 1–10 million years, quickly enough that all of the atmospheric methane should have been destroyed by now. The presence of even a trace amount of methane in the atmosphere means that it must be resupplied continually to the atmosphere. One possibility for resupplying the methane is the evaporation of methane from the surface. Since methane on the Titan surface would be in the form of a liquid rather than solid ice at the surface temperature of 94 K, the possibility exists that there might be an ocean or lakes of methane. It would be evaporation of methane from the liquid at the surface that could resupply the atmosphere with methane. If enough methane could build up to saturate the atmosphere, then methane could condense into clouds.

We can explore the possibility of the existence of a methane ocean or lakes by looking at whether the atmosphere is, in fact, saturated with methane. To do this, we will use the Earth's atmosphere as an analogy. If the Earth's atmosphere was composed of nitrogen and oxygen only, with no water vapor, there would be no clouds and the temperature would be expected to change within the lower atmosphere by 10 K for every kilometer change in altitude. This rate of change, called the 'lapse rate' of the atmosphere, defines the temperature profile that would result from vertical mixing of a dry atmosphere – as air is raised to higher altitudes it cools off at the rate of 10 K/km, and as air is pushed to lower altitudes it heats up at the same rate. With water vapor nearly saturating the Earth's atmosphere, however, cooling the gas will cause some of the water to condense out as liquid; as it condenses, it will give off energy (the so-called latent heat), and this energy will keep the atmosphere from cooling off as rapidly. In a 'wet' atmosphere, then, the lapse rate will be only 6.5 K/km. In fact, this is the average observed temperature change with altitude in the lowermost part of the atmosphere. The existence of widespread oceans on the Earth provides a ready source of water to keep the lower atmosphere saturated.

On Titan, methane might serve the equivalent role to the Earth's water vapor. A nitrogen atmosphere saturated with methane would be expected to have a lapse rate of about 0.3 K/km. This is much less than the observed lapse rate, however, which means that the methane abundance must be less than would be required to saturate the atmosphere. Thus, although methane is an observed constituent of the atmosphere, the atmosphere cannot be fully saturated and the clouds cannot consist of methane. This implies that there cannot be a global or nearly global ocean of methane covering the surface; if there were, the methane would easily evaporate into the atmosphere, which would become saturated with methane, and the temperature lapse rate would match the predicted value.

Still, there could be a global ocean if it did not consist entirely of pure methane. For example, if liquid ethane (C_2H_6) were mixed into the ocean, it would lower the vapor pressure of the methane; this would reduce the amount of methane that could evaporate into the atmosphere. An ocean that was about 70% ethane, 25% methane, and 5% nitrogen could exist and be in equilibrium with the atmospheric constituents; it would also be consistent with the temperature measurements. Is this composition plausible?

When methane (CH_4) absorbs solar ultraviolet light, it breaks down into CH_2 and CH. These combine together to form the molecules C_2H_4 and C_2H_2, with the release of some additional hydrogen gas (as H or H_2). The hydrogen gas will escape to space quickly because it is so light, so these longer-chain hydrocarbon molecules cannot recombine to form methane again, and the conversion is irreversible. C_2H_2 acts as a catalyst, in fact, allowing the sunlight to convert methane into ethane. As ethane gas is produced, it eventually will saturate the Titan atmosphere and begin to rain out onto the surface; there, it would exist as a liquid. Estimates suggest that all of the methane in the atmosphere could be consumed and converted to ethane in only 1–10 million years. If these same processes operated continuously for the history of the solar system, 4.5 billion years, an amount of ethane would be produced that would form a liquid layer on the surface that was about a kilometer thick. It would seem that a global ethane ocean containing some methane is certainly possible. In addition, the C_2H_2 and other organic molecules that are produced could be responsible for the thick hazes in the atmosphere.

Are there any measurements that could tell us whether a global ocean actually exists on Titan? In fact, there are two different approaches, but they both provide results suggesting that a global ocean probably does *not* exist.

The first approach involves making radar observations of Titan from the Earth. Radio waves are transmitted toward Titan from a large radio telescope located in the California desert. They bounce off of whatever materials are at the surface of Titan, and some of the energy bounces back toward the Earth. There, it is received by the Very Large Array (VLA) of radio receivers located outside of Socorro, New Mexico. The VLA is the largest array of radio telescopes on the Earth, and is the most sensitive set of instruments for this type of measurement.

Two different characteristics of the received radio waves are important. The first is the amount of energy reflected from the surface. If Titan had a surface that consisted of either an ethane–methane ocean or a silicate rock material similar to the surface of the planets in the inner solar system, it would reflect about 10% of the radio energy hitting it, with the rest being absorbed by the surface. Not all of the reflected energy would be reflected back toward the Earth, but we can tell how much total energy is reflected from how much is actually measured at the receiving antenna. In fact, much more than 10% of the energy is reflected. The amount reflected is consistent with a surface that is predominantly composed of ice – not smooth ice, but a rough, jagged surface where radio energy is efficiently scattered a few times before heading back toward the Earth. This same type of reflection is seen when radio waves are bounced off of icy planets in the solar system, such as the icy satellites of Jupiter (Europa, Ganymede, and Callisto), the polar ice caps on Mars, and the polar ice caps that were discovered by using this technique on Mercury. Some liquid methane or

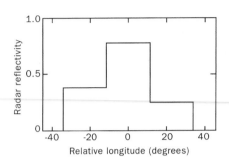

[Figure 12.4]
Radar reflectivity of Titan as a function of longitude. The relatively high values compared with that expected for an ethane/methane mixture, combined with the substantial variability from place to place, suggest that a global ocean is not present. (After Muhleman *et al.*, 1990.)

ethane may be present on the surface and still be consistent with the radar results, but a global ocean cannot be present.

The second characteristic of the radar measurements is that they can be used to provide crude maps of the surface of Titan. These maps show variations from one place to another in the amount of radar energy that is reflected back to the Earth (Figure 12.4). This variability means that the reflectivity of the surface is not the same everywhere. If there were a global ocean, we would expect it to be uniform and to have the same radar reflectivity everywhere. Again, a global ocean does not appear to be likely.

The second approach involves obtaining images of Titan at wavelengths in the near infrared, just longward of the wavelengths to which the human eye is sensitive. At these wavelengths, the clouds in the atmosphere are much more transparent than they are at visible wavelengths, and we can see through to the surface. (There is still some absorption of light by the atmosphere, so images show surface features in a way similar to looking through a fog – they can be seen, but not as clearly as on a clear day.) Images of Titan were obtained in 1994 using the Hubble Space Telescope. Using a telescope in orbit around the Earth allows smaller features to be viewed than could be seen using a telescope at the surface of the Earth; features as small as 300 km were observed. These images showed variations from one place to another in the amount of sunlight that was reflected (Figure 12.5). Large, continent-sized bright patches were seen, and they appeared to be fixed on the Titan surface. Again, the existence of spatial variations in the reflectivity of the surface argues against the presence of a global, uniform ocean.

How can we reconcile the theories that predict the formation of a kilometer-thick, global ocean of ethane and methane with the observations that suggest that a global ocean is not present? It is not an acceptable solution to suggest that there is no ongoing production of ethane from methane in the atmosphere. There clearly is methane in the atmosphere,

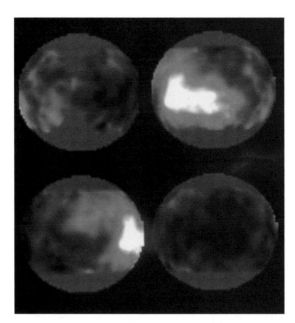

[Figure 12.5]
Hubble Space Telescope images of Titan. At this near-infrared wavelength, some sunlight penetrates to the surface and is reflected back to space. Thus, the variations seen in the images are indicative of variations in the albedo of the surface. (NASA photo.)

and relatively straightforward calculations show that it will be chemically processed to produce ethane. In addition, the spectral measurements from the *Voyager* spacecraft indicate the presence of a wide variety of organic molecules in the atmosphere, with methane being the most likely starting point. The carbon cycle on Titan must involve evaporation of methane from the surface into the atmosphere, dissociation of atmospheric methane by ultraviolet sunlight, production of ethane and carbon-rich organic hazes in the atmosphere, and condensation of the ethane onto the surface as a liquid (Figure 12.6).

One possibility, however, is that the ethane–methane liquid mixture does not all reside right at the surface as open bodies of liquid. Rather than being an ocean, the liquid might be able to percolate into the subsurface through cracks and pores in the surface ice, creating the Titan equivalent of terrestrial groundwater. If the ice at the Titan surface has a porosity that is typical of the surface regions of other planets and satellites, then a porous surface layer that is ten kilometers thick might be able to hold the equivalent of a kilometer-thick layer of liquid. (For comparison, the Mars regolith is estimated to be able to hold the equivalent of up to about 1.5 km of water in its subsurface. The Earth's land regions hold an average of about a 50-m layer of groundwater, but are capable of holding much more than this.)

A second possible solution involves the uncertainty in our estimate of the ethane production rate. The atmospheric temperature profile and the

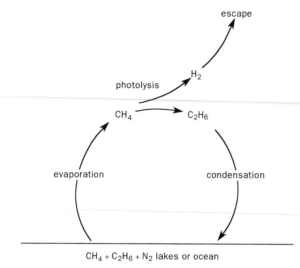

[Figure 12.6]
Schematic diagram showing the evolution of methane into higher-order hydrocarbons. Methane evaporates into the atmosphere from surface lakes or oceans. There, ultraviolet light from the Sun breaks the methane apart, the components recombine into ethane, acetylene, and propane (only ethane is shown), and any remaining hydrogen escapes to space. The hydrocarbons produced in this process saturate the atmosphere, condense into liquid, and rain back onto the surface. (After Lunine, 1985.)

composition are not known exactly, and a variety of different values can be used that still are consistent with the various measurements. Calculations using these different numbers result in different photochemical production rates for ethane, primarily because the rates at which photochemical processes occur are temperature dependent. Using the alternative rates can reduce the estimate of the amount of ethane produced over time to the equivalent of a thickness less than 300 m. Such a lower amount of ethane could be accommodated on Titan even at the surface without necessarily having a global ocean. If some of the ethane indeed does infiltrate into the subsurface, then there is no difficulty hiding much or all of this from view. Remember, of course, that some methane and ethane liquid must be at the surface in order for the methane to be able to evaporate and replenish the atmospheric methane. However, because the methane is removed from the atmosphere on timescales of 1–10 million years, the replenishment of the present atmosphere could have occurred any time during the last several million years; it does not necessarily have to be continuing at the present.

Water on Titan

As we begin to look at the possible relevance of Titan to our understanding of exobiology, and at the specific questions concerning the chemistry of

Titan, one remaining issue is the existence of water on Titan. Because of its great abundance in the universe, we expect a substantial fraction of the mass of Titan to consist of water. However, the low surface temperature, around 94 K, means that the water would be present as ice rather than liquid. Can there have been liquid water on Titan at some point in its history?

The most likely way in which liquid water could occur would be if there had been some heating event that was able to melt the water ice. It is unlikely that a planet as small as Titan could have produced enough heat on its own (by the decay of radioactive elements, for instance) to have melted the water ice near the surface. However, heat could have been supplied by the energy released in the impact of a large asteroid onto the surface. This could have occurred either during the formation of Titan, when energy from the impacts of the accreting planetesimals would be available to melt the water, or during Titan's subsequent history, with impacts occurring throughout geologic time.

For an object with Titan's mass and radius, the escape velocity is 2.6 km/s; this also is the velocity that an object would obtain just from falling through Titan's gravitational field. Of course, an object that collides with Titan must first have fallen through part of Saturn's gravitational field; falling through Saturn's gravitational field to the location of Titan's orbit would give an object an additional velocity of 7.9 km/s. When it collides with Titan, it can be moving either in the same direction as Titan in its orbit or in the opposite direction; that is, it can collide head-on or from the rear (or, of course, anywhere in between). And, a colliding object will have had its own initial velocity relative to both Saturn and Titan before even approaching the Saturn system. Thus, an object hitting Titan could have had an impact velocity anywhere between about 3.5 and 30 km/s. At the larger impact velocities, enough heat can be generated by the impact to melt a substantial amount of material at the surface. In fact, it has been estimated that the impacts that have occurred throughout time would have provided enough heat so that every part of the Titan surface would have been melted at least once. Given the length of time required to freeze again, a typical piece of the surface would have contained liquid water for 100–1000 years. That is, there appears to have been enough heat to melt water ice at least occasionally.

The presence of liquid water as at least a transient species has implications for prebiotic chemistry or even for biology.

It has another interesting ramification as well. During a localized or global melting event due to an impact, we would expect that any rocky material that was present would settle to the bottom of the liquid water.

204 This would deplete the surface of rock, and we would expect that the surface should consist either entirely or almost entirely of ice. Of course, either mixed in with the ice or occasionally sitting on top of it would be the mixture of liquid methane and ethane that was described earlier.

Titan exobiology

Let us now examine the Titan environment from the exobiology point of view. This will help us to understand whether Titan has ongoing chemistry that would be analogous to terrestrial prebiotic chemistry or possibly even whether it might have any active biota.

Despite the probable existence of substantial amounts of water ice at the Titan surface and in the subsurface, there probably is no liquid water on the Titan surface today. The only possible exception to this might be short-lived, transient 'reservoirs' located near the site of any recent impact events. Titan probably is too small to have any significant interior heating at the present due to the decay of radioactive elements; this means that vents of hot volcanic material or liquid water, analogous to terrestrial hydrothermal systems near recent volcanic activity, are not likely to be present today.

Liquid methane and ethane almost certainly are present at the surface of Titan and possibly in the subsurface, either as large-scale (but not global) oceans, lakes, or 'groundwater'. However, these are not likely to lead to biological activity. Both methane and ethane are relatively non-reactive chemically. They do not have the property that water has of being able to dissolve a wide variety of compounds. Organic molecules that may form in the Titan atmosphere will probably settle to the bottom of the ocean or lake rather than remain dissolved or suspended as a prebiotic soup.

There is one other fundamental difference between liquid water and liquid ethane and methane as a medium in which life could exist. Water interacts quite easily with silicate minerals. It does so both physically (the way in which rivers can carve valleys in solid rock) and chemically. Chemically, liquid water will interact with typical minerals to form clay minerals. As discussed in Chapter 6, clay minerals on the early Earth may have provided chemically active sites that helped in the formation of life. The possible absence of clays on Titan would make this possible mechanism for the formation of life extremely unlikely.

Not only do methane and ethane not interact substantially with silicate minerals, but there probably are no substantial quantities of silicates available at the Titan surface. The large amount of ice present on Titan probably resides at the surface, with the silicates and any iron having settled into a central core, as described above. As a result, access to the biogenic elements other than hydrogen, oxygen, carbon, and nitrogen would be extremely limited. The only source of silicon, calcium, iron and phosphorous would come from meteorites or asteroids that would have impacted onto the surface subsequent to the formation of a Titan core. We see this same dearth of biogenic elements at the surfaces of the other icy satellites in the solar system, such as the Jupiter satellites Europa, Ganymede, and Callisto. These objects also have substantial quantities of water ice. Their surfaces are, in fact, dominated by ice, with only a small fraction of non-icy material.

Despite the lack of ubiquitous liquid water and the dearth of many of the biogenic elements, a substantial amount of atmospheric organic chemistry that is related to terrestrial prebiotic chemistry is occurring on Titan. The atmosphere consists primarily of nitrogen and methane. Some ethane is present in the atmosphere, and the liquid at the surface (whether it be in the form of an ocean, a lake, or groundwater) probably consists of a mixture of these species. Chemical reactions in the atmosphere are driven by ultraviolet light from the Sun and by high-energy charged particles from the Saturn magnetosphere. Both of these are able to break the chemical bonds in CH_4 and N_2 (although the amount of energy required to break the N_2 bond is substantial so that nitrogen chemistry will be relatively less important). The breaking of these bonds allows other carbon- or nitrogen-containing molecules to form. We expect to see production of such organic molecules as C_2H_6, C_2H_2, HCN, and HC_3N, for example, and each of these species has been observed in the Titan atmosphere (Figure 12.7). The organic molecules produced in the atmosphere will saturate the atmosphere and produce an atmospheric haze; in fact, such a haze contributes to the clouds that obscure the Titan surface from view.

As the organic molecules saturate the atmosphere and form clouds, some of the cloud material will settle out of the atmosphere. This will produce a layer of organic debris within the methane and ethane oceans and lakes, as well as organic sediment on the surface. Over the age of the solar system, over 100 m of sediment should have been formed. The sediment is estimated to consist of the equivalent of a layer of C_2H_2 that is 100 m thick, 10 m of HCN, 1 m of more complex nitrogen compounds, and 1–10 m of solid, more complicated organic molecules. In addition, the heat associated with the occasional impact events will form ponds consisting of a slurry of water, methane, ammonia, and organic compounds. Such a

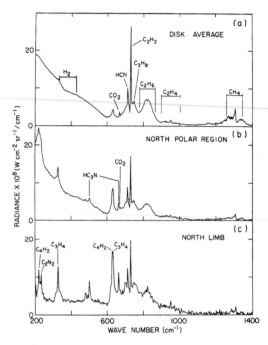

[Figure 12.7]
Thermal–infrared emission spectra of the Titan atmosphere, taken from the IRIS experiment on the *Voyager* spacecraft. The abscissa is in 'wave numbers', and corresponds to emission at wavelengths from 50 to about 7 micrometers. Each spectrum shows emission features that are characteristic of various organic molecules. (From Samuelson *et al.*, 1983.)

slurry is very similar to what might have been present on the prebiotic Earth, although perhaps containing an even higher concentration of organics. Chemical reactions will occur in this slurry, and amino acids and organic polymers might result. These are just the organic precursors that are thought to have led to the formation of life on Earth. Although the length of time required for the formation of life on Earth is uncertain, most people think that it is unlikely that life could originate on Titan in the short time span during which liquid water might be present locally.

Concluding comments

On Titan, then, we see a planet on which substantial organic chemistry has occurred and is thought to be ongoing at the present. Much of this chemistry is similar to that which might have occurred on the early Earth and might have led to the formation of life. This similarity is especially intriguing if the production of terrestrial prebiotic organic molecules occurred by atmospheric processes as was suggested by Miller and Urey (see Chapter 6). On the other hand, if terrestrial life arose out of organic molecules produced in hydrothermal systems or brought in by incoming extraterrestrial debris, then the processes on Titan would have less relevance. And, if the origin of life actually occurred at hydrothermal systems, then it is not clear that Titan has any clear relevance to the formation of terrestrial life. Despite this, however, Titan appears to have clear exobiological relevance. It is the best natural laboratory that we have today for exploring possible prebiotic terrestrial chemistry in a planetary environment. Understanding the chemical reactions that occur on Titan and their organic byproducts will be of substantial benefit to understanding the early Earth. And, in conjunction with a search for organic molecules and possible prebiotic chemistry or biological activity on Mars, Titan will help us to understand the processes that occurred on the Earth and led to the formation of life.

A spacecraft mission to Saturn and to Titan was successfully launched in 1997, and carries a probe that will enter the Titan atmosphere and land on the surface. The orbiter portion of the spacecraft, known as the *Cassini* orbiter, has instruments on it that can measure the surface properties of Titan; these include a radar mapper to map out the radar reflectivity and look for oceans or lakes of methane and ethane. The probe that will enter the Titan atmosphere, known as the *Huygens* probe, will measure the atmospheric temperature and composition; it includes a mass spectrometer that will look for the organic constituents of the atmosphere. And, of course, when the probe hits the surface, it will find out whether there is an ocean. If it lands on a liquid surface, it will float and send back measurements of the properties of the liquid. And if it lands on a solid surface, it should be able to image the surface. The exploration of Titan is just beginning.

be very small and could not contribute substantially to the overall lunar mass and density. Importantly, recent analysis of the *Galileo* measurements of the Io gravity field indicates that it is denser toward the center than at the surface. This supports the existence of a core. Of course, there is no way to know whether an increased density from having additional iron could be balanced by a decreased density from having some ice; Io could contain some ice along with a dense core and still have the observed density. However, there is no evidence for any water ice at the surface or water vapor (or its byproducts oxygen and hydrogen) in the atmosphere.

Europa's average density is about 3.01 g/cm^3. If it consists of typical rock material made of silicate minerals with some water ice added, this density requires that about 10–15% of the mass of Europa consist of ice. At the distance of Europa from the Sun, 5.2 A.U., temperatures are warm enough that water is the only ice that is expected to be present; other ices would sublimate to space relatively quickly, and they might never have been able to condense at these relatively high temperatures. The surface of Europa is highly reflective, and the wavelength behavior of the reflected sunlight suggests that the surface is composed essentially of pure water ice. There must be some contaminating debris mixed in, though, because some of the surface areas are a little bit darker than pure ice would be and they have a 'dirty' appearance. Measurements from the Galileo spacecraft suggest that Europa has a center core that is made of a denser material. As a result, the water ice abundance actually has to be a little more than 15%, in order to give the observed average density.

Ganymede and Callisto, the largest two of the Galilean satellites and the farthest from Jupiter, both have densities below 2.00 g/cm^3; Ganymede's is 1.93 and Callisto's is 1.83. As with Titan, these values are substantially lower than expected for a pure silicate composition, and they require a substantial amount of ice to be mixed in. Ganymede's density requires that between about 50 and 57% of the mass consist of ice, and Callisto's requires that between 52 and 58% of the mass be ice. The uncertainty in determining an exact ice abundance, both for these two and for Europa, comes from the uncertain distribution of the ice within the satellites. Because the density of ice depends on the pressure that is being applied, the exact amount of ice required will be different if all of the ice is at the surface than if some of the ice is buried deeply. Also, some of the water could have combined with silicate minerals to produce hydrated minerals; these have lower densities than non-hydrated silicates and would contribute to a lower overall density without requiring pure ice as a mixer.

What makes the two satellites Io and Europa of interest from the possible biological viewpoint is the additional input of heat into their interiors through tidal heating. The tidal heating on Io is somewhat analogous to the raising of tides on the Earth by the Moon. Here, the gravitational force from the Moon pulls on the Earth's oceans. As a result, the water flows through the oceans toward the point that lies beneath the Moon, and it accumulates to a greater thickness there (and on the opposite side of the Earth as well, due to the differential pull of the Moon and the rotation of the Earth). Because the Earth rotates on its axis once every day, we see this as a rising and falling of the sea level twice each day. The movement of the water is stopped by the presence of the continents and by friction within the ocean itself, and these processes dissipate the energy of the moving water. On the Earth, the amount of heat released by this friction is negligible, so we do not see a noticeable increase in the surface temperature.

On Io, the tides are created by the movement of Io in its orbit around Jupiter. The orbit is not exactly circular, so Io is closer to Jupiter and then farther away from it at different times during each orbit. As it moves in and out, the force from Jupiter's gravitational pull changes slightly, and the changing pull distorts the shape of Io. In addition, because the orbit is not exactly circular but Io rotates on its own axis at a constant rate, Io would appear to wobble back and forth a little bit as seen from the surface of Jupiter. As it wobbles, the point on the Io surface that is closest to Jupiter changes slightly. Because the gravitational pull from Jupiter pulls on Io and distorts its shape, this wobble also causes the shape of Io to be altered a little bit each orbit. Just as friction that results from rapidly bending a paper clip back and forth will heat up the paper clip, distortion of Io's shape with each orbit will dissipate energy and cause Io to heat up. This frictional dissipation of the tidal energy puts enough heat into the interior of Io that much or all of its interior has been heated to the melting point and active volcanism is produced at the surface. And, in fact, the entire surface appears to be volcanic! Active volcanic plumes have been imaged by the Voyager and Galileo spacecraft, and lava lakes and volcanic landforms are seen on the surface (Figure 13.2).

The energy released as heat in the Io interior must come from somewhere. For instance, the tidal energy that is dissipated in the Earth by the Moon actually comes from the rotation of the Earth and from the energy in the Moon's orbit. As heat is released, the Earth is slowing down its rotation and the Moon is moving farther away. In the past, the Moon was much closer to the Earth, and the Earth rotated much more quickly. Four billion years ago, the Moon may have been only a few thousand kilometers away from the Earth, and an Earth day would have lasted only about five hours. On Io, the energy comes from the orbit of Io itself. As the energy dissipates, its orbit around Jupiter should become circular very quickly. If the orbit

were circular, then the shape of Io would not change as it went through its orbit, and no tidal dissipation or tidal heating would exist. The orbit does not become circular, though, because the gravitational tug from nearby Europa pulls on Io and pumps up Io's orbital eccentricity (the degree to which the orbit is elliptical rather than circular). This happens the same way that you can pump up a child on a swing by pushing him once each swing – Io and Europa are locked in a resonance that causes Europa to pull on Io at regular intervals. As a result, some of the energy is coming from Europa's orbit as well. Ganymede participates in this little dance as well, pushing on Europa's orbit. Callisto is the only Galilean satellite that does not play in this game.

Theoretical models of this dissipation of energy suggest that a very large amount of energy is dissipated within the Io interior. This much heating will warm up the interior, and some of the heat will be lost by conduction to the surface or by thermal radiation from surface volcanic eruptions. In addition, there is some heating of Europa's interior, although much less than within Io.

Io

What's the connection of the Io volcanism to possible life? As we have discussed for the Earth and Mars in particular, the presence of hydrothermal systems associated with volcanism may have played a special role in the formation and evolution of life. The heat from the volcanism provides a source of energy that can be tapped by living things. The heat is not accessed directly, however, but is utilized through its influence on chemical activity. If Io has active volcanism and a source of geothermal energy, could it harbor life?

The most amazing feature of Io is its abundant active volcanism. Essentially the entire surface of Io is covered by volcanic material, and active volcanism has been observed both from spacecraft and through Earth-based telescopes. There is a wide variety in the types of volcanic eruptions that occur on Io, although not as wide a variety as on the Earth. There are features that may have solidified from molten lava lakes (Figure 13.3). There are individual lava flows and central-vent volcanoes (Figure 13.4). Volcanic landforms that appear to be similar to the shield volcanoes of the Earth's Hawaiian islands are apparent (Figure 13.5). However, there are no large volcanoes analogous to the largest and most explosive volcanoes on the Earth such as Mt. St. Helens. One feature is observed on Io that

[Figure 13.2]
Image of the surface of Io taken from the *Voyager* spacecraft. Essentially everything that can be seen in this image was formed by volcanic processes, with different types of eruptions leaving different features on the surface. (NASA photo.)

[Figure 13.3]
Image of the volcanic vent
Loki on Io. It may have been
a molten lake at one time,
but observations suggest
that it is not molten at
present. (NASA photo.)

is not seen on Earth – plumes of volcanic gas and dust that are ejected up to a couple hundred kilometers above the surface (Figure 13.6). These plumes appear to be similar to terrestrial geysers, although on Io the fluid that drives the eruption is not water but sulfur dioxide, SO_2.

Most of the color seen on Io probably results from the presence of sulfur in various forms. SO_2 frost is white (which accounts for the white patches surrounding volcanic vents), and sulfur itself can exhibit a wide variety of brown, yellow, red, and orange colors. Much of the volcanism on Io is thought to be sulfur volcanism, produced by melting various sulfur compounds rather than by melting silicate rock. There must be some silicate volcanism on Io as well, given the presence of some features that cannot be due to sulfur, although the proportion of silicate volcanism versus sulfur volcanism is not known.

Although active volcanism occurs on Io, there is no evidence that any water exists. In fact, it is not clear whether Io ever had any significant amounts of water. Io is close enough to Jupiter that heat received from Jupiter during the satellite's formation may have kept the temperatures high enough that water would not have condensed. There would have been some scattering of planetesimals between the Galilean satellites during their formation, though, so some water-rich objects from farther out in the Jupiter system may have accreted onto Io. Even if there was water on Io at some point, however, it may not have been able to remain there for very long. The active volcanism would have kept the surface warm near the

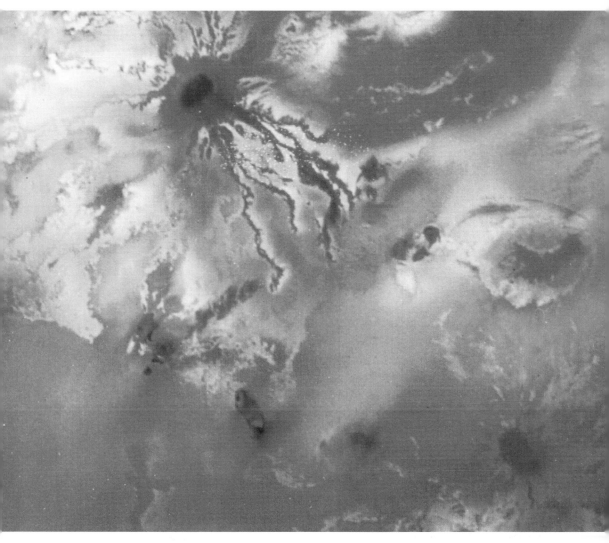

[Figure 13.4]
Volcanic vent on the Io surface, showing lava flows on the surface and their source region. (NASA photo.)

volcanic eruptions. The higher temperatures would have caused any water frost or ice at the surface to evaporate into the atmosphere; there, it could have escaped to space relatively easily because of Io's low gravity. And, the plumes themselves would have spewed any water into space directly. The high degree of volcanic activity would mean that every location on the surface probably would have been heated at least occasionally.

There is no evidence for the presence of water on Io in any form – this includes the surface, the atmosphere, and the region of space near Io. Could there be hidden water? It is possible that, locally, active volcanic

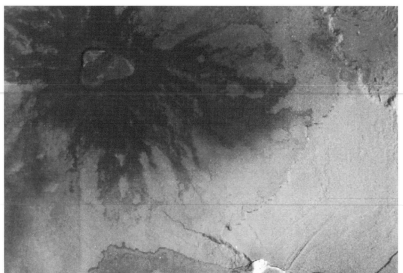

[Figure 13.5]
A 'Hawaiian-style' volcanic caldera (collapse pit), surrounded by flows of lava over the surface. (NASA photo.)

sites might contain small amounts of water that is just now being released from deep within the interior. If a dehydrating planet contained sites that possessed both active volcanism and small amounts of water, exobiological activity could possibly exist. If there is life on Io, however, it would have to be hidden away in these locations. This does seem like a stretch, however. Life would have the additional problem of trying to survive in the incredibly strong radiation environment at Io; the energetic particles trapped by the Jupiter magnetic field would be very damaging to living organisms. The most reasonable conclusion is that, despite the active volcanism, there is neither water nor life on Io.

Europa

Europa, the next moon out from Io, presents a different story, however. Europa probably contains about 15% or more water, based on its density. If the water is distributed uniformly throughout the total volume of Europa's interior, it could exist in the form of hydrated silicates, minerals that contain water bound into their structure. These are relatively common on Earth, and include such minerals as serpentine and clays; they form by chemical interactions between water and volcanic silicate minerals, for example. Alternatively, if Europa has differentiated into layers, the water

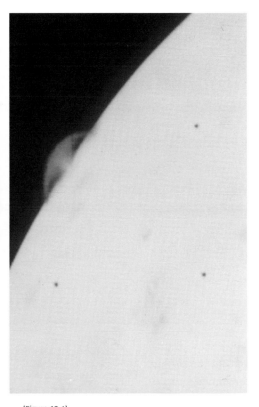

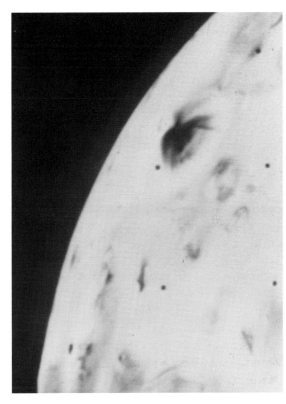

[Figure 13.6]
Volcanic plume on Io, shown rising above the surface into the atmosphere. The 'mushroom' shape is typical for particles traveling on ballistic trajectories. (NASA photo.)

would have been released from these minerals and would rise to the surface. There could be a layer of water that is up to 300 km thick at the surface of Europa that, as we will discuss, could be melted at relatively shallow depths; this could produce an ice-covered ocean of liquid water that conceivably could harbor life.

Although there is no direct evidence whether or not the water within Europa forms a layer that is concentrated at the surface, we do know that there is abundant water ice at the surface. We see evidence for this in the absorption of characteristic wavelengths of the sunlight reflected from the surface. In fact, the ice appears to be relatively pure, containing few contaminants.

In addition, the dark material that does contaminate the surface of Europa gives the surface a 'cracked' appearance (Figure 13.7). A large number of linear features are present, which give the surface a broken appearance that may result from the application of tectonic stress at the surface (Figure 13.8). Many of the cracks are filled with what looks like a very clean water ice, suggesting that water ice might be present beneath the surface as well (Figure 13.9). High-resolution images taken from the Galileo

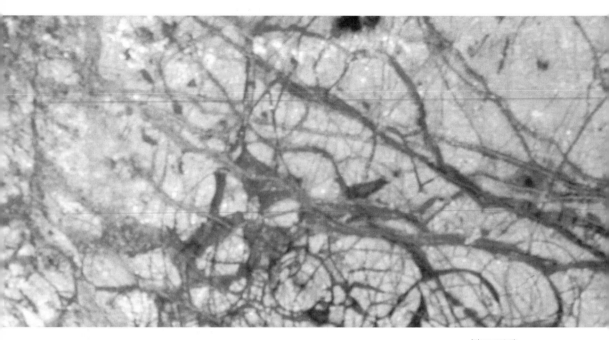

[Figure 13.7]
Surface of Europa, imaged by the *Galileo* spacecraft. The cracked appearance of the surface shows up here as dark bands separating lighter 'plates'. (NASA photo.)

spacecraft show clear evidence for material having moved over the surface (Figure 13.10). If this movement was not caused by liquid water erupting to the surface from a buried ocean, and blocks of surface ice floating across the surface, then it must have resulted from the flow of ice. And, the ice would need to be warm, near the melting temperature, to flow so readily.

The very young age of the Europa surface, seen from the small number of impact craters that are present, requires there to have been some substantial resurfacing of the planet in recent times. This resurfacing, again, could have taken the form of eruptions of liquid water to the surface or flows of soft water ice.

If there is a thick layer of water ice at the surface, could there be liquid water beneath the surface, forming an ice-covered ocean? To answer this question, we need to look at the heat balance of Europa's interior. The same tidal forces that heat up the interior of Io also heat the interior of Europa, but with much less intensity; Europa also will be heated by the decay of radioactive elements within the rocky part of the interior. As the interior heats up, the temperature difference between the surface and the interior will increase. When there is a temperature difference, heat will conduct from the warmer to the cooler region. Therefore, as the interior begins to heat up, the heat will be conducted from the interior to the surface where it can be radiated away to space. The temperature of the planet eventually

[Figure 13.8]
In this close-up image of the Europa surface, the dark bands are seen to consist of a series of parallel ridges, as if the light ice was being pulled apart and darker ice was injected from beneath to fill the gaps. Pairs of 'grooves' criss-cross the surface. (NASA photo.)

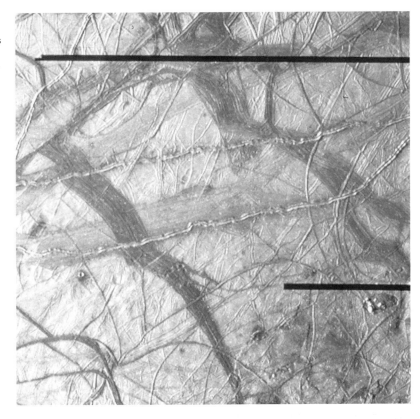

[Figure 13.8]
In this close-up image of the Europa surface, the dark bands are seen to consist of a series of parallel ridges, as if the light ice was being pulled apart and darker ice was injected from beneath to fill the gaps. Pairs of 'grooves' criss-cross the surface. (NASA photo.)

will reach a steady state, neither heating up nor cooling off, when as much heat is being conducted to the surface and lost to space as is being dissipated by tidal heating and produced by radioactive decay.

We can estimate the amount of tidal and radioactive heating, and therefore how much the interior heats up below the surface. This will allow us to calculate how deep into the ice we would have to dig before the temperature would be high enough that the water ice would melt.

If Europa is covered with a thick ice layer, then the amount of tidal heat dissipated within its interior is about 8×10^{17} erg/s, or about 2 erg/cm^2-s. About twice as much heat would be dissipated if there is a layer of liquid water beneath the surface, and perhaps twice as much again if there is a molten layer within the core. The amount of heat produced by radioactive decay probably is not more than is generated inside of the Earth's Moon; the heat flux from the Moon today is about 30 erg/cm^2-s. In order for this heat to make its way to the surface by conduction, the temperature beneath the surface would need to increase by as much as 0.005 K for every

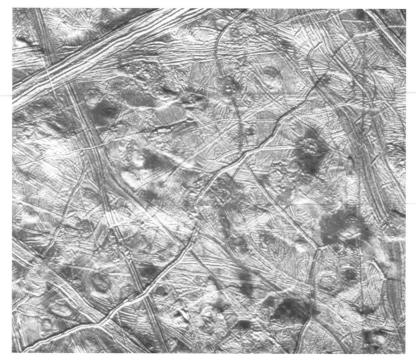

[Figure 13.9]
Criss-crossing grooves on the
Europa surface. The grooves
indicate compression and
extension of the ice.
(NASA photo.)

meter of depth, or 5 K for every kilometer (assuming that the surface is
solid water ice). With an average surface temperature of about 120 K and a
melting temperature for water ice of 273 K, the ice could be warm enough
to melt at depths of only a few tens of kilometers. If the layer of ice is only
about 100 km thick, however, and if the total heating is at the low end of the
range of possible values, then the temperature at the bottom may not be
warm enough for the ice to melt.

There are possible factors that might change this conclusion, however.
First, the ice near the surface may not be dense water ice. A porous frost does
not conduct heat as easily as solid ice, so the temperature gradient would need
to be larger in order for the heat dissipated within the interior to get out. If the
surface consists of a very porous, fine-grained frost, the temperature could
change by as much as 5 K for every meter of depth; this is 1000 times greater
than the value mentioned above. This would mean that the melting temperature of ice could be reached at depths of only a couple hundred meters, and
that most of the mantle of water could be liquid. The surface layer probably is
very porous, with the ice becoming more consolidated at deeper layers as
compaction and heating solidify it. Thus, it is possible that a liquid ocean
might be encountered at any depth below a couple hundred meters.

[Figure 13.10]
Close-up of the Europa surface. In this image, the main mass of ice appears to have broken up into blocks, and the blocks have moved across the surface. Individual grooves can be traced from the main mass to the blocks, and the amount of movement can be seen. The blocks give the appearance of having floated away on a water ocean that subsequently froze, or on a layer of warm, soft ice; either way, the break-up of the main block and the movement of the fragments indicate a high likelihood of there being liquid water near to the surface. (NASA photo.)

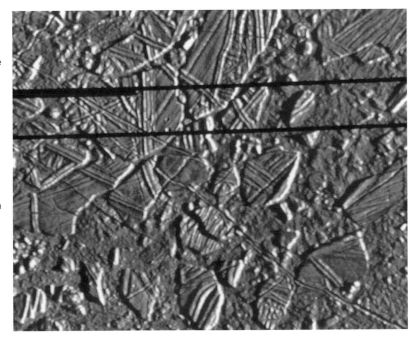

Second, there is a phenomenon that may increase the temperature at a depth of only a few meters by a substantial amount over the typical surface temperature of 120 K. If this occurs, then water could melt at even shallower depths than were predicted above. The phenomenon involves the penetration of sunlight below the surface instead of being absorbed right at the surface. This could occur because ice is transparent to sunlight, and the sunlight could penetrate many meters below the surface before being absorbed. If it does this, then the depth at which the sunlight is absorbed would heat up to temperatures well above the surface temperature. This behavior would be similar to that which occurs in a greenhouse, where sunlight enters easily but heat does not escape as easily. If Europa has such an 'ice greenhouse' at its surface, then the temperature at only a few meters depth could be as much as 100 K warmer than the surface, and ice could melt at very shallow depths. However, surface temperature measurements made from the *Voyager* spacecraft showed a substantial day-to-night variation on Europa, which appears to rule out the penetration of sunlight to more than a couple of centimeters below the surface. With sunlight penetrating only to these very shallow depths, the additional ice greenhouse warming can be no more than about 10 K, rather than the possible 100 K. Therefore, there probably is not a substantial ice greenhouse warming of Europa.

222   Given these factors, there is a strong likelihood that liquid water exists at some depth beneath the surface of Europa. And, even if there is no global buried ocean, the tidal heating could be localized and produce local, small-scale pockets of liquid water – perhaps buried lakes. Although the images of the surface show intriguing evidence that suggests the presence of liquid water at least intermittently, there is no proof that water exists.

If there is liquid water, can there be life? In addition to water, we need a source of energy to drive biological activity and the presence of the biogenic elements. Both of these could be present at the bottom of a Europan ocean, where the water would be in contact with the silicate minerals that lie below.

If the minerals comprising the silicate portion of Europa are similar to those that occur elsewhere in the solar system, then we expect easy access to the biogenic elements at the ocean–silicate boundary. And, there is no reason to expect that they would be different on Europa.

The energy present in liquid water at a temperature near the melting point cannot by itself drive biological activity. The ability to form organic molecules or to drive biochemical reactions comes from chemical disequilibrium. On Titan, we saw that ultraviolet light from the Sun could drive the disequilibrium, but we do not expect sunlight to penetrate to the depths where liquid water could exist on Europa. On the early Earth, it is possible that lightning could have triggered the disequilibrium, and that the resulting chemical reactions could form organics and then possibly life; again, though, we would not expect lightning deep beneath the Europa surface.

What we might expect, however, is that the usual planetary geological processes would occur within the silicate portion of the satellite. Europa is only a little bit smaller and less massive than the Earth's Moon (about 10% smaller in radius, 1569 versus 1738 km, and 40% smaller in mass, 4.5×10^{25} versus 7.4×10^{25} g). Because it is smaller and less massive, the silicate portion of Europa, which contains some radioactive elements within the individual minerals, should generate less internal heat from radioactive decay than the Moon. But, the tidal heating of Europa would make up this difference, so we might expect the silicate portion of Europa to be similar in its level of geologic activity to the Moon. The Moon was volcanically active during about the first 1.5 billion years of its existence; volcanic eruptions were common then, with lava that erupted to the surface filling large impact basins and creating the dark mare regions that are seen on the side of the Moon that faces the Earth. Volcanism on Europa, beneath an ocean of water, might resemble terrestrial volcanism that occurs at mid-ocean spreading centers; there, lava is emplaced onto the surface forming a series of 'pillow' basalts and associated intrusive volcanism.

And, there might be hydrothermal systems at the boundary between the ocean and the 'seafloor' that would be similar to those on the Earth's seafloor. The chemical disequilibrium that results from changing the temperature of the water as it circulates through the hot volcanic rock could drive the production of organic molecules and possibly of life. This process could be similar to the possible origin of life in the terrestrial hydrothermal systems. Similarly, weathering of the primary volcanic rock by the water could release energy that could be utilized by living organisms; similar sources of energy are tapped by some bacteria on Earth and might be available on Mars as well (see Chapter 8).

As we did with the energy sources on Mars (Chapter 8), we can estimate the global inventory of energy sources on Europa. If we take the amount of volcanism at the surface of the rocky portion of Europa to be similar to that on the Moon (see Figure 7.3), we can estimate the total volcanic energy available through hydrothermal systems. And, by looking at possible chemical weathering of rocks at the water–rock interface, we can estimate the energy from that source. The total amounts of energy are quite low – volcanic energy would be much less than was available on Mars, and the weathering energy will depend on how thick a layer of rock might be able to weather. Whether there was sufficient energy available to power an origin of life, or whether there might be life today (or possibly evidence of past life within the water or ice layer) is uncertain.

Is life on Europa plausible? We really won't know until we explore it further, but all of the necessary ingredients may be present. One way to search for biochemical or biological activity without having to drill through an unknown thickness of ice to the ocean below would be to look at the debris deposits that are present on the Europan surface. Where water or ice may have flowed through the cracks in the ice to the surface, the deposits on the surface may include organic material carried there from beneath the surface.

Life in the Jupiter atmosphere

What about Jupiter itself? It seems bizarre even to consider the possibility of life on Jupiter. Jupiter is a gas-giant planet, consisting of material that has nearly the same overall composition as the Sun. The parts of the planet that are visible to the naked eye or in telescopes are the cloud decks in the atmosphere. In fact, the atmosphere is thought to extend to extremely great depths, and a surface of the sort that we think of as being present on a planet – made of rocky material – may exist only in the very deepest parts of the interior. Furthermore, the interior of Jupiter is too hot to allow organic molecules to exist; they would break apart into their constituent atoms. If there is any life on Jupiter, it would have to be confined to the part of the atmosphere near the visible cloud layers, where temperatures are comparable to those in the Earth's atmosphere, for example. However, even if there is no life, there certainly is interesting organic chemistry going on there.

The Jupiter atmosphere consists primarily of hydrogen and helium, which are present in solar abundances. In that sense, Jupiter is similar to the Sun. All of the other elements that are present in the Sun are present on Jupiter as well, but as trace constituents. The major biogenic elements other than hydrogen – O, C, N, and S – all have been detected. C, N, and S are present in very nearly the same relative abundances as in the Sun, although they are present combined with other atoms, in the form of CH_4, NH_3, and H_2S. Oxygen exists, within H_2O, but it is substantially depleted in the visible part of the atmosphere relative to its abundance in the Sun.

In order to understand the possibilities for Jovian life, we first need to understand a little bit about the chemistry and energetics of the atmosphere. The atmosphere is, overall, extremely turbulent. It is overturning constantly in order to get rid of excess heat derived from the gravitational contraction and settling of the interior. Because of this overturning, combined with a rapid rotational period (length of day) of only about 10 hours, the atmosphere has stratified into distinct regions of upwelling gas and regions of downwelling gas. These regions, the 'zones' and 'belts', respectively, form bands that circle the planet at constant latitude (see Figures 1.5 and 13.1). The zones contain gas that cools off as it rises in the atmosphere, with some of the gases condensing into clouds that probably consist of condensed H_2O, NH_3, and NH_4SH. When these gases condense, the resulting clouds all would be white, though, and the Jupiter clouds are various shades of brown, red, and blue. Thus, there must be additional coloring agents, or chromophores.

These chromophores might include phosphorous compounds such as phosphine, PH_3, which has been detected in the Jupiter atmosphere, various sulfur molecules, or organic molecules. Simple organic molecules have been detected in the atmosphere, including methane (CH_4), ethane (C_2H_6), acetylene (C_2H_2), and hydrogen cyanide (HCN). If the Jovian atmosphere were in chemical equilibrium, all of the carbon would be present as methane; in addition, all of the nitrogen would be in the form of NH_3, and the oxygen would be contained in H_2O. Since there is carbon, oxygen, and nitrogen present in these other molecules, processes are occurring in the Jupiter atmosphere that push the system away from chemical equilibrium. These processes may be driven by the energy from sunlight, may involve the mixing of gases up to the level of the clouds from deeper layers where different molecules are stable, or may involve biological processes. Sunlight certainly plays a role, as it is capable of breaking apart the methane molecules in the upper atmosphere and creating some of the other organic molecules that are seen.

The processes that form more-complex organic molecules in the Jupiter atmosphere could be some of the same ones that may have operated on the early Earth (see Chapter 6); lightning and sunlight may have played significant roles. Complex organic molecules have been produced in laboratory simulations of these processes under Jovian conditions; essentially, the same types of molecules produced in the Miller–Urey discharge experiment can be produced in the Jupiter atmosphere. Without a doubt, there is organic chemistry of exobiological interest occurring on Jupiter.

What about life? The biggest difficulty that life would face on Jupiter is that any large, macroscopic biota would tend to sink down into the deep atmosphere, and any small, microscopic bacteria would be carried downward in the downwelling regions of the atmosphere. As organisms are carried to deeper levels, they would encounter higher temperatures that would destroy organic molecules. Any life that might exist could survive only in the cooler upper parts of the atmosphere. It is possible, however, that Jovian biota could get around this problem: they could multiply quickly enough for a large number of offspring to be produced before the parent is destroyed. Although many of the offspring would be pulled into the deep atmosphere and killed, some of them might be raised up to the higher altitudes where they could live, thereby allowing survival of the species. Or, they could create gas-filled sacs that would provide some buoyancy in the Jupiter atmosphere; this would allow them to survive at temperate altitudes without sinking too deeply into the atmosphere. However, a creature could grow large enough to have sufficient buoyancy only if it had some mechanism of powered locomotion that would allow it to ingest other creatures and use their mass as construction materials for new growth – that is, if it could hunt. Such creatures conceivably could grow large enough to be visible in spacecraft images.

Even though life on Jupiter may be possible, it seems somewhat implausible. Although the proper sized bacterium might be able to multiply and survive as a species on the proper timescale, the environmental conditions that would allow survival are very narrow and would place very stringent requirements on a species. It would be difficult for a species to evolve to have the proper characteristics without first passing through stages at which survival would not be expected to occur. For example, an origin of life on Jupiter, analogous to the origin on Earth (Chapter 6), might have involved the formation of organic molecules and some simple repro-ducing species that was much smaller than modern terrestrial species. These molecules and simple species would be small enough to be caught in whatever motions the atmosphere was undergoing. This would include being carried relatively rapidly into the deep layers where organic mole-cules would decompose. Survival of simple organisms seems very difficult.

In addition, the origin of life on Earth may have taken place in a prebi-otic 'soup' in liquid water. Such a soup existing in the Jupiter atmosphere would be extremely thin – the organic molecules present today, and the biogenic elements necessary for life to exist, are present in very minute quantities and would be extremely difficult to access.

On the other hand, we have to keep in mind how little we really under-stand about terrestrial life. We are finding species living in environments on Earth that only ten years ago were thought to be inimical to life. On Earth, living species seem to occupy almost every conceivable niche suit-able for their existence. Might the same occur on other planets, despite the difficulty of life getting to that niche?

Concluding comments

Is life in the Jupiter system possible? Although we have no evidence what-soever to suggest that life might actually be present, there are reasons to think that life could exist there. If one lists the major criteria that seem to be required for life – liquid water, a usable source of energy, and access to the biogenic elements – there are possibilities here. Io has at least two of the three (energy and the biogenic elements) and could have the third (water); however, there is no direct or indirect evidence for the presence of water on Io, and there are reasons to think it might be absent. Europa has two of the three (energy and the biogenic elements) and may have the third; a mantle of liquid water could exist on Europa beneath an ice cover-ing although, again, the presence of liquid water, although suggested, is not required by any existing measurement. Jupiter also has two of the three (energy and liquid water, with the liquid water occurring as clouds at some altitudes within the atmosphere) but it seems unlikely that the biogenic elements are present in sufficient quantities to allow life.

After Earth and Mars, Europa would seem to be the most likely place in our solar system on which (or, possibly, *in* which) to look for life. It may meet all three environmental requirements for the occurrence of life. The next most likely planets would include those in which at least two of the three requirements are met – Io and Jupiter in this system, and Titan in orbit around Saturn. In order to possibly find abundant planets on which life could occur, however, we may have to venture outside of our own solar system.

228   The possibility of life in our solar system is limited to the few planets that seem to have the necessary prerequisites for life, the most limiting of which appears to be the existence of liquid water. As we expand our search for life out into the universe, we need to understand the likelihood of finding planets around other stars. We are especially interested in those planets that are Earth-like in nature – those which might have liquid water on their surfaces and which might therefore be able to harbor life. Although astronomers are both searching for and finding planets around other stars (Chapter 15), we do not yet have the ability to find Earth-like planets. As a result, we have to use the understanding that we have developed of planet formation to try to predict whether Earth-like planets can exist.

Our goal in this chapter is to look at the processes that were responsible for the formation of the planets in our solar system and to see how these same processes might operate around other stars. We will start with our own solar system, because we know the most about it and can best understand the processes responsible for its formation. In fact, it is the only solar system for which we really know what properties need to be explained by any theory for planet formation. Once we have a sense of what processes operated here, we can look to see whether these same processes might be widespread throughout the galaxy and what types of solar systems might have resulted. In the next chapter, we will look at the observational evidence for the existence of planets around other stars and in Chapter 16 we will discuss whether they might be habitable.

Important characteristics of our solar system

There are nine planets in our solar system, of widely varying sizes and compositions. These include the rocky planets in the inner solar system, the gas-giant planets in the outer solar system, and Pluto, which appears to be in a category by itself. Of course, planetary satellites can be considered to be worlds in their own right, based on the occurrence of diverse geological processes at their surfaces that have been driven by internal processes. Taking this into consideration, the list of planets should be expanded to include at least our own Moon, the Galilean satellites Io, Europa, Ganymede, and Callisto, the Saturn satellite Titan, and the Neptune satellite Triton; this list could easily be expanded further until we have over 20 planets. In order to understand how these objects might have formed, and whether similar objects could have formed around other stars, we need to

look at the characteristics of the planets individually and of the ensemble
of planets as a whole. A tremendous amount of order and regularity in the
solar system exists, and the nature of this regularity provides telling clues
as to the formation processes.

We begin then with a list of the major characteristics of the planets and
of the solar system. It is difficult to know ahead of time which of these will
be important general properties that might apply to other solar systems
and which are specific to ours. However, this will lead into a discussion of
the implications that these properties have for the early history of our solar
system and for its formation.

1. All of the planets in our solar system orbit around the Sun in the same
   direction, and this is the same as the direction of the Sun's own rotation. In
   addition, most of the planets also rotate on their own polar axes in this same
   direction; the exceptions are Venus, which rotates in the opposite direction,
   and Uranus and Pluto, which are tilted somewhat on their sides.
2. The orbits of all of the planets are in nearly the same plane, and this plane is
   aligned very closely with the Sun's equator. Pluto's orbit is tilted the most, by
   about 20 degrees with respect to the orbits of the other planets. In addition, the
   orbits of the nine planets are nearly circular; again, Pluto's orbit departs from
   circular the most, as given by its orbital eccentricity, but even so not by very
   much.
3. To the best of our knowledge, all of the planets formed at about the same time,
   around 4.5–4.6 billion years ago. Admittedly, we have direct determinations of
   age for only a few objects – the Earth, the Moon, Mars (through the martian
   meteorites, as discussed in Chapter 8), and the meteorites from the asteroid
   belt – but, all of these are very similar. The age of the Sun cannot be measured
   directly, but estimates based on observations of stars of similar mass in the
   galaxy are consistent with an age of 4.5 b.y.
4. The planets are well spaced throughout the solar system, with orbits that do
   not cross each other. Neptune and Pluto are the sole exception to this rule.
   Pluto's orbit departs enough from circular that at its closest it can be closer
   to the Sun than Neptune. Neptune and Pluto are in a complicated resonance,
   however, so that whenever Pluto is close to the Sun Neptune is on the opposite
   side of the Sun; they never approach each other very closely. The spacing of the
   orbits of the planets can be described by what is known as the Titus–Bode law –
   the sequence of numbers 4, 4+3, 4+(3×2), 4+(3×4), 4+(3×8), 4+(3×16), and
   so on, divided by 10, accurately yields the distances of the planets from the Sun
   in astronomical units (A.U., where 1 A.U. is defined as the average distance of
   the Earth from the Sun). This law seems to work quite well except for Neptune,
   which might look like an intruder into our solar system, and the asteroid belt,
   which exists between the orbits of Mars and Jupiter where the Titus–Bode law
   predicts the existence of a planet. Although regularly spaced planets might be
   a natural consequence of the formation process, it is unlikely that the details
   of this particular law are anything other than a random occurrence.
5. The compositions of the nine planets differ from each other, in a way that
   seems to relate to their distances from the Sun. The four planets closest to the

Sun – Mercury, Venus, Earth, and Mars – are all rocky planets with thin (or no) atmospheres. The next four planets – Jupiter, Saturn, Uranus, and Neptune – are the gas giants, much larger planets with very thick atmospheres and without solid surfaces in the familiar sense (although there might be rocky cores at their center). Pluto, the farthest planet from the Sun, is a small, icy object, in many ways more like one of the icy satellites in the outer solar system than like any of the other planets.

6. The atmospheres of Jupiter and Saturn are dominated by hydrogen and helium, the main constituents of the Sun and of the universe; their overall composition is very similar to that of the Sun. In addition, those meteorites thought to be the least altered since their formation 4.5 b.y.a. also have a composition similar to that of the Sun (except for hydrogen and helium, which are heavily depleted in the meteorites). Uranus and Neptune do not have as much hydrogen and helium as Jupiter and Saturn, and their atmospheres appear to be dominated by molecules normally thought of as ices or volatile compounds – water, carbon dioxide, methane, and ammonia.

7. Planetary satellites are common, but not ubiquitous. The gas giants all have miniature 'solar systems' of satellites, with most satellites going around their central planet in the same direction as the planet's rotation and nearly in the planet's equatorial plane. All of the gas giants also have rings consisting of a very large number of particles, from micron-size dust up to house-size blocks, orbiting around them. The inner planets have neither rings nor an ensemble of satellites. Earth has its Moon, but Venus has none; Mars has two very small satellites, but these may be intruders from the asteroid belt.

8. There is a cloud of comets around the Sun, at a distance of about 10^4–10^5 A.U. Although this cloud cannot be seen, its existence has been confirmed based on the most distant point found in orbits derived for long-period comets that pass through our solar system. Comets in this cloud, named the Oort cloud after Jan Oort, the astronomer who first predicted its existence in the 1950s, can be perturbed by passing stars, causing some of them to pass through the inner part of the solar system. Based on the orbits taken by these comets as they pass through the inner solar system, they appear to be distributed uniformly in a spherical cloud, rather than being confined to the same plane as the planets.

9. There is a second cloud of comet-like objects, located past Neptune and Pluto, at a distance of 30–50 A.U. These comets, however, are located in a belt or a disk that is in the same plane as the planets, and they orbit the Sun in the same direction. This belt of comets, called the Kuiper belt after the astronomer Gerard Kuiper, is thought to be the source region for short-period comets. Kuiper-belt objects were discovered observationally in the 1990s, and more than 30 are known today; they range in size from 100 to 400 km, and there probably are a very large number of smaller objects. (Pluto may turn out to be simply the largest of these objects.)

10. Added together, the planets are a very small fraction of the total mass of the solar system, with the Sun containing more than 99.8% of the mass. Jupiter contains most of the rest of the mass. On the other hand, the planets, and Jupiter in particular, contain most of the angular momentum of the solar system. Angular momentum is the tendency toward rotation. Each planet, with the rotation from its orbit around the Sun, contributes to the total angular

momentum of the solar system. Even though the Sun rotates on its axis once
every 27 days and thereby contributes to the total angular momentum, the
large distance of Jupiter from the Sun causes the less-massive Jupiter to
contribute more.

11. Meteorites date back to about 4.5 b.y. ago, and their chemical properties
    suggest that temperatures existing prior to their formation were above 1500 K,
    at least in the inner part of the solar system. At this high temperature,
    substantial vaporization of any solid material would occur. They also constrain
    the time that it took for the material to evolve from a distributed cloud of
    material to solid grains, some of which survive almost unaltered up to the
    present; this time was less than 10 m.y.

Since the various parts of the solar system are confined to a disk out to
well past the orbit of Pluto and almost everything revolves and rotates in
the same direction, it has been suggested that the solar system originated
as a distributed disk of material that accumulated together to form the
individual planets. The common ages of the individual components and
the common composition of the various materials leads to the conclusion
that the Sun and this 'protoplanetary disk' originated from a relatively
(although not completely) homogeneous cloud of gas and dust. If it had
even a small original rotation, the collapse of such a giant cloud would
have led naturally to a flattened disk orbiting around a central mass. Most
of the mass in this system would have ended up in the central mass, a
protostar which became the Sun. The rest would have been left behind in
this collapse and would have formed a disk of gas and dust revolving
around the proto-Sun. This debris then would have accumulated to form
the planets. This 'solar nebula' origin for the planets was first suggested by
Immanuel Kant in 1755, based on many of these same observations, and
has been embraced by the scientific community as being consistent with a
variety of observational and theoretical results.

Formation of 'protoplanetary' disks around stars

Our solar system is thought to have formed from the collapse of a giant
cloud of molecular gas and debris. Such clouds are seen in many parts of
our galaxy, and young stars are common within them. The universe is
thought to be about 15 billion years old, and our galaxy is probably
comparable in age. Why do stars continue to form? The life cycle of stars is
not a simple one, with all stars having the same history from the time of
formation of the galaxy. The lifetime of a star depends strongly on its mass.
Massive stars, ten or more times the mass of the Sun, burn their nuclear

fuel rapidly and have short lifetimes. Less massive stars, those about 0.1 times the Sun's mass, burn their fuel at a much slower rate and last much longer. The lifetimes of individual stars range from only tens of millions of years up to longer than ten billion years. Thus, while some of the stars that we see today may be first generation, having been created at about the same time as the galaxy, the majority of stars that formed early have gone through their complete life cycles. New stars continue to be created out of collapsing gas clouds, however, and we see stars in our galaxy that have a wide distribution of ages.

The most massive stars use up their nuclear fuel quickly and begin to collapse. The collapse, in fact, triggers a rebound that results in a massive, sudden expansion of the star and the ejection of a large portion of its mass to space. This process is a supernova explosion of the star, and the debris shed by the explosion gets redistributed throughout the nearby region of interstellar space. In addition, stars that do not explode can shed some of their matter back into the galaxy through their 'stellar winds'. Because only hydrogen, helium, and lithium, the three lightest elements, formed during the Big-Bang origin of the universe, it is within stars themselves that all elements heavier than these were produced. And, it is through supernova explosions and stellar winds that these elements, along with the heaviest elements that are produced only in the explosions, are redistributed throughout the galaxy. The fact that our Sun and planets contain these heavy elements means that the material in our solar system once resided inside of stars.

The shock wave from the expansion of clouds of gas and dust from a supernova can trigger the collapse of a nearby giant molecular cloud. In order for collapse to occur, the density of the cloud has to be increased to a high enough value that the gravitational pull of the cloud will overcome the tendency of the gas to expand. As the cloud collapses, it may break up into a number of smaller clouds, each of which will collapse independently to form one or more stars. This collapse takes a very short time (around 10^5 years).

As this collapse occurs, two important things happen. First, any rotation that the cloud initially has will be preserved as it collapses. As a result, the collapsing cloud will begin to rotate faster in the same way that a figure skater can spin faster by pulling her arms in closer to her body. As a result of the spinning, the cloud cannot collapse completely to form a single spherical ball that contains all of the cloud material; it would be spinning too fast and would break apart. Instead, the cloud collapses to a central accumulation of material (destined to become the central star) and a surrounding disk of gas and dust. Although the rotation keeps the material from moving

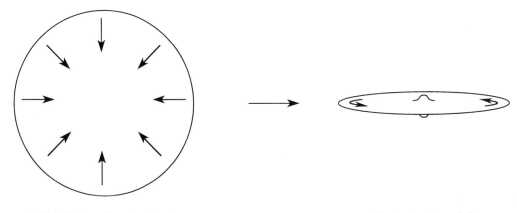

Initial giant molecular cloud
~1-10 light years (~10^{18}-10^{19}cm) across

Protoplanetary disk
~100 A.U. (1.5x10^{15} cm) across

[Figure 14.1]
Schematic diagram showing the gravitational collapse of a giant molecular cloud of gas and dust to form
a protoplanetary disk. Because the initial cloud is turbulent, it will have some net rotation; as a result of
the conservation of angular momentum, the collapse will form a rotating disk rather than a sphere,
with a protostar embedded in the center.

closer to the axis of rotation, it does not keep it from collapsing parallel to
this axis (keeping the same distance from the rotation axis); as a result, the
gas and dust will collapse to form a disk of material rotating around the
central mass. The result is the presence of a disk with a composition very
similar to that of the protostar, including the full abundance of hydrogen
and helium (Figure 14.1).

Second, as the material in the inner part of the disk collapses, it is com-
pressed to higher densities and, because of the increased friction, heated
up. At distances from the protostar corresponding to the inner part of our
solar system, temperatures become high enough to vaporize much of the
dust. As the disk cools off, recondensation can occur. Hydrogen and
helium, along with some of the other gases, are never able to condense.
Water probably did not condense within the inner solar system. As
condensation proceeds, we were left with a disk of gas and dust-sized
debris. It is from this debris that the planets eventually form.

The gaseous component of the solar nebula is not present today, so
some process capable of removing it must have occurred. We see such a
removal process occurring around a type of young star named a 'T Tauri'
star (named after the first example that was found). There, gas flows
outward from the surface of the star, in a manner similar to the solar wind
flowing out of the Sun today but in much greater amounts. This outflow of
gas is able to drag along and remove any gas and dust left behind after the

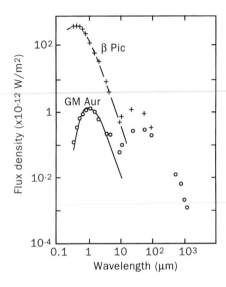

[Figure 14.2]
Observed flux from the stars Beta Pictoris (β Pic) and GM Aurigae (GM Aur). In each case, the flux at wavelengths below about 10 micrometers is fit well by a blackbody function that matches the emission expected from a star. At longer wavelengths, there is an 'infrared excess' that corresponds to thermal emission from a disk of dust surrounding the stars. From the amount of emission, the average disk temperatures must be in the vicinity of 100 K, corresponding to disks with sizes comparable to that from which our own solar system formed. (After Beckwith and Sargent, 1996.)

formation of the planets. The fact that hydrogen and helium are present in Jupiter and Saturn in nearly their solar abundances means that these planets must have formed early enough to trap these gases while they still were present. However, the dearth of H and He in the atmospheres of Uranus and Neptune suggests that much of the gas had already been removed from the solar system prior to their formation.

Is there any evidence that such protoplanetary disks actually can form around young stars? Indirect evidence for the existence of these disks comes from observations from the Infrared Astronomical Satellite (IRAS) that orbited the Earth in the early 1980s. A large number of stars were observed to have an excess in infrared emission at wavelengths that correspond to heat energy (Figure 14.2). From the wavelengths at which this emission occurred, the temperature of the objects that were doing the emitting was inferred to be about 100–400 K. This emission is thought to come from a large ensemble of individual particles in orbit around the young stars. Planets themselves would be too small to produce as much emission as is observed, but the same amount of mass distributed as dust grains throughout a large disk could do so. The temperatures correspond to disks ranging up to 100 A.U. in size; these disks are comparable in size to our solar system and to the nebula out of which our solar system is thought to have formed.

Direct evidence for the presence of these disks comes from telescopic images obtained for a small number of these stars. When the light emitted by the star itself can be masked out, so as not to blind the camera, starlight reflected from the surrounding disk can then be seen. As one can imagine,

[Figure 14.3]
HST images of two protoplanetary disks. In each case, the disk is dark due to absorption of light reflected off of nearby gas clouds. The disks appear elongated because they are seen somewhat edge-on. Each disk is about 100 A.U. across, roughly the size of our own solar system. (a) HST 16. (b) HST SW; notice the cloud of gas that appears to be in the process of being blown away by a strong stellar wind. (NASA images courtesy of J. Bally and D. Devine.)

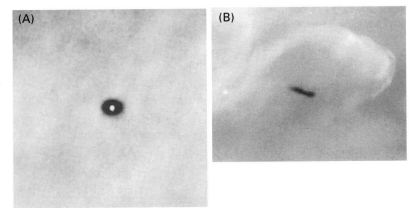

this is a difficult observation to make because the star itself is so much brighter than the disk. Images have been taken of only a small number of disks and from a few stars close to our Sun. In the case of the best-imaged disk, around the star Beta Pictoris, some of the detailed structure in the disk can be seen as well (see Figure 1.8). In a couple of instances, although the light reflected from the disk cannot itself be seen, the disk can be seen silhouetted against a bright gas cloud behind it (Figure 14.3).

The best astronomical evidence suggests that between about one quarter and one half of the young stars in our galaxy have dust disks surrounding them. Since this dust can accumulate to form planets, it means that planets might be common!

The accretion of planets

How do planets form from the disk of gas and dust that surrounds the protostar? The process described here has been worked out for our own solar system, but it is thought to be typical of what might happen in a protoplanetary disk around any young star. It involves the accumulation of dust to form objects that are 1–10 km in size, termed planetesimals, followed by the gravitational accumulation of these planetesimals into full-sized planets. The outer planets start with a rocky core formed by this accretion process, but are also able to accumulate gas from the solar nebula before the gas disperses.

Let's start with a discussion of how the process occurs in the inner part of the solar system, where the terrestrial planets will reside.

The dust grains in the cooling protoplanetary disk would be too small for there to be any significant gravitational attraction between them. Rather, dust grains will stick together due to electrostatic forces or to 'contact welding' when they collide. The turbulent movement of the gas and dust in orbit around the proto-Sun ensures that the particles will collide with each other, allowing growth to occur.

Imagine that the material that eventually will form the terrestrial planets was initially distributed in a disk of dust that occupies the part of the solar system where the inner planets are today. Given the current mass of the planets, the resulting disk would have had a mass density of about 10 g/cm^2. This is an area density, meaning that every square centimeter on the face of the disk would have a total of 10 grams of dust sitting above or below it. The dust would not all be exactly in the plane of the disk, but would occupy a finite volume both above and below the disk. With this much dust present, collisions would occur often enough that planetesimals in the size range of 1–10 km could accumulate in only 10^5 years. At the end of this time, most of the dust would have resided in planetesimals in orbit around the Sun.

Once they reach the 1–10 km size range, the planetesimals are large enough that each one will begin to exert a gravitational pull on the adjacent ones. As each planetesimal can pull in other nearby planetesimals, collisions will occur at a greater rate. By this process of collision of kilometer-sized objects, planetesimals can continue to grow.

Once the planetesimals are able to attract one another gravitationally, accretion into larger bodies proceeds quickly. The nature of the growth depends on whether there is much gas left in the inner solar system (as the gas will slow solid particles down and keep impact velocities lower) and on whether the largest impacts might instead lead to break-up of the planetesimals. The most likely result is that the largest objects will grow the fastest, due to their greater ability to gravitationally attract and accrete more planetesimals. This leads to what has been termed 'runaway' accretion, in which the largest object in a given region of the solar system accumulates most of the mass (it 'runs away' from the rest in terms of its size and mass). The accretion by the largest protoplanet of such a large number of smaller objects causes its orbital properties to average out – it ends up in a nearly circular orbit and accumulates all of the nearby mass.

At the end of this runaway phase of accretion, which takes about 10^6 years, most of the mass in the inner solar system is contained in objects that are Moon to Mars sized (1 to 10% of the mass of the Earth), and these objects are in nearly circular orbits that are, on average, about 0.02 A.U. apart. There are thought to have been hundreds of these 'planetary embryos' in the inner solar system.

Once they become Moon sized or larger and contain most of the mass in the solar system, these planetary embryos will be able to gravitationally perturb one another's orbits. Very quickly, the orbits will become more eccentric (less circular), and orbits of adjacent embryos will begin to cross. When this occurs, collisions between embryos can occur. These collisions result in the final growth of the planets and in the removal (or sweeping up) of much of the remaining debris from the solar system. Also, close encounters between embryos, without a collision occurring, will perturb the embryos into even more eccentric orbits. The eccentricities of the orbits can become high enough for substantial mixing to occur between the embryos formed throughout the inner solar system. That is, each planet does not form only from planetary embryos in its region of the solar system – there is no such thing as a discrete 'feeding zone' for each planet. Rather, planetary embryos from the location of Mercury's orbit to that of Mars and the asteroid belt will contribute to each planet. Mixing is not complete, though, and there will still be a slight preference for embryos in each region to accumulate together.

This final stage of planetary accretion is complete only when there are no more planetary embryos to sweep up. Numerical computer simulations of this process indicate that it will take around 10^7–10^8 years. What is most interesting is that a large number of simulations suggests that, typically, three or four planets would be expected to have formed in the inner part of the solar system. Individual simulations usually include one planet with a final mass very similar to that of the Earth, one planet with a slightly smaller mass, like that of Venus, and one or two additional, smaller planets. The simulations suggest that 'Venus' can be either inside or outside of the orbit of the Earth, and that 'Mercury' and 'Mars' may or may not be present. These simulations closely resemble our own solar system and suggest that our solar system may be typical of other solar systems that are out there (Figure 14.4).

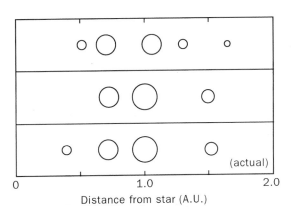

[Figure 14.4]
Results of simulations of the formation of planets in a protoplanetary disk. The top two panels show the planets that resulted from a single simulation each, obtained using initial conditions that differ randomly. The bottom panel shows the planets in the inner part of our own solar system. (After Wetherill, 1996.)

(actual)

0   1.0   2.0

Distance from star (A.U.)

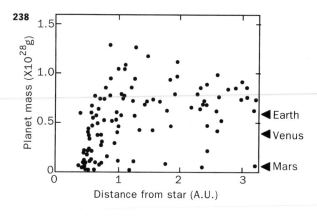

[Figure 14.5]
Results of simulations of the formation of planets in a protoplanetary disk around a Sun-like star. The results from 33 different simulations (with different initial conditions) are shown together, in order to give a sense of where planets can form and how massive they would be. For each planet formed, the final mass and the distance from the central star are shown; the masses of Earth, Venus, and Mars are indicated for reference. In these simulations, no effects have been included due to the presence of Jupiter. (After Wetherill, 1991.)

The most likely place for the most massive planet to occur was determined to be at about 1 A.U. from the Sun – the location of the Earth (Figure 14.5). It would seem again that our solar system is typical of what we might expect! In fact, simulations for the formation of planets from disks around other stars suggest that solar systems that form there should be very much like ours. Terrestrial-like planets appear to be a natural consequence or byproduct of the formation of stars, and they should be widespread throughout the galaxy.

What about the asteroid belt in our solar system? The computer simulations suggest that the presence of Jupiter perturbs the orbits of the planetesimals and planetary embryos in this part of the solar system. Some of these objects can be thrown into the inner solar system, where they can be accreted by the protoplanets there, and some can be ejected from the solar system entirely. The perturbations from Jupiter continue at the present, causing the eccentricities of the asteroid orbits to be greater than they otherwise would be and resulting in collisions between asteroids being at higher velocities than would otherwise occur. These collisions cause the asteroids to break apart, preventing the accumulation into larger objects. The asteroids cannot accrete into a single planet and appear to be destined to remain as a very large number of smaller objects. If Jupiter had not been present, it is likely that the debris in the asteroid belt would have accumulated to form one or more additional rocky planets.

An interesting result of the final accretion of the planets being from Moon- and Mars-sized embryos is that giant collisions would be a necessary consequence during this phase. The final planets formed by the catastrophic impact of objects that were themselves planet sized. Interestingly, giant impacts were postulated to have occurred even before this theory of accretion was developed. During the 1980s, it was realized that the best explanation for the origin of the Moon was for a Mars-sized object to have collided with the growing Earth. The Moon then would have formed by the accumulation in orbit around the Earth of debris that was ejected into space by the impact. This scenario was best able to explain the composition of the Moon along with its orbit and other characteristics. In addition, a giant impact was hypothesized to have occurred on Mercury. If the impact occurred after the planet's core had been segregated from the mantle, then much of the silicate mantle could have been ejected into space; this would explain Mercury's relatively high average density. In this case, the debris did not form a satellite but instead must have been lost to space.

An obvious question to ask at this point is why the Earth has a moon but Venus does not. Both planets were subjected to similar accretion processes and giant impacts, but only in one case did it result in the ejection of material capable of forming a moon. We currently have no answer, although there is no requirement that all impacts and all planets behave similarly.

This scenario of accretion by impact also is consistent with what we see recorded in the geological record on the Moon. By the time the largest planetary embryos had been swept up and the planets could begin to record their history on their now-solid surfaces, most of the large debris in the solar system already would have been accreted onto one of the planets. Only the smaller debris was left behind for a final sweeping-up process. Of course, this debris included objects hundreds of kilometers across that were able to form, for example, the largest and oldest impact basins on the Moon (Figure 14.6). Gradually, this material also was swept up. Today, there still is debris that is colliding with the Earth occasionally (see Chapter 2); most of it, however, originated in the asteroid belt and was only recently perturbed into Earth-crossing orbits.

Understanding the formation of the gas giant planets in the outer solar system requires a modification to this process of accretion. In addition to rocky material, the outer planets must have accreted some of the gas from the solar nebula as well. As mentioned, Jupiter and Saturn, with abundances of most elements in their atmospheres matching those in the Sun, would have to have accreted a large amount of gas. Neptune and Uranus, on the other hand, which show a substantial depletion in hydrogen and helium compared with the Sun, must have accreted much less gas from the nebula. The problem for Jupiter and Saturn is that gas from the solar nebula does not become gravitationally bound onto growing protoplanets until they are about 10 or 20 times the mass of the Earth. Thus, the formation of Jupiter and Saturn must have started with the formation of rocky

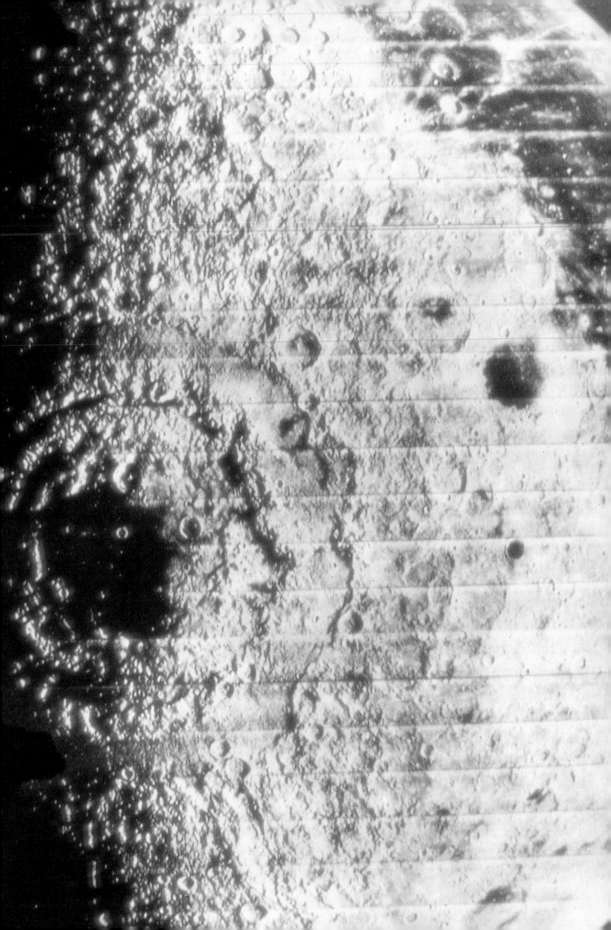

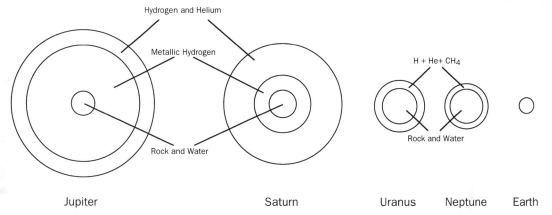

Hydrogen and Helium

Metallic Hydrogen

H + He+ CH$_4$

Rock and Water

Rock and Water

Jupiter   Saturn   Uranus   Neptune   Earth

[Figure 14.7]
The interior structure of Jupiter, Saturn, Uranus, and Neptune. The planets' sizes are shown to scale, along with the Earth for comparison. Each of the four planets has a rocky core, surrounded by hydrogen- and helium-rich material. Uranus' and Neptune's outer layers are depleted in hydrogen and helium relative to their abundances in the Sun and in the atmospheres of Jupiter and Saturn, although H and He still are the most abundant atmospheric species.

cores by an accretion process similar to that which occurred in the inner solar system. This accretion would have formed protoplanets that were 10–20 times the mass of the Earth, at which point they would have begun to trap gas as well.

Observations of the 'T Tauri' outflow of gas from young stars in our galaxy suggests that the gaseous nebula surrounding a young star is lost within 10 million years after the star forms, although this period is very uncertain. The rocky cores of Jupiter and Saturn must have formed in less time than this, so that there still would have been gas in the nebula that could be accreted onto the planets. Uranus and Neptune, being farther out in the solar system, would have to have taken longer to form a rocky core massive enough to begin to attract gas from the nebula; as a result, they would have accumulated much less gas by the time the T Tauri phase of the Sun removed the remaining gas.

Models of the interior structure of the outer planets suggest that all four planets have a high-density core at their centers that is 10–20 Earth masses in size. This is the same size as the rocky core that is required to begin to accumulate gas. This similarity probably is not coincidental. The fact that Jupiter is 318 times as massive as the Earth means that it is composed primarily of the gas accreted onto this rocky core; Saturn, at 95 times the Earth's mass, also is composed primarily of solar-nebula gas. Uranus and Neptune, on the other hand, at 14.5 and 17.1 times the mass of the Earth, respectively, must have accreted only a small portion of nebular gas on top of their rocky cores (Figure 14.7).

[Figure 14.6]
Lunar Orbiter mosaic of the Orientale impact basin on the Moon. The outer ring is about 900 km across. The horizontal 'stripes' are an artifact of the process of making the mosaic.

Finally, consider the Oort cloud of comets at about 10^4 A.U. Comets could not have formed easily at such large distances from the Sun, owing both to the low density of the distant solar nebula and to the low orbital

242 velocities and long times between encounters with dust or debris. They must have formed closer to the Sun and then been ejected to the outermost reaches of the solar system. The current thinking is that comets were formed as ice-rich planetesimals somewhere near the current orbits of Uranus and Neptune. As these planets accreted, close encounters of planetesimals with them would have perturbed the planetesimals' orbits. While some were ejected from the solar system entirely or sent into the inner solar system, most were sent into orbits that carried them out past 1000 A.U. There, galactic tides or encounters with nearby molecular clouds would further perturb the orbits. The velocity added from these perturbations would have been about the same as the orbital velocity at that distance from the Sun, which would effectively 'randomize' the orbits. As a result, we see only a random cloud of comets today. Additional perturbations occurring much later could send some of the objects back into the inner solar system, where they are seen as comets.

The Kuiper belt, on the other hand, would have to consist of objects that formed farther out than the orbits of Uranus and Neptune. There, the dust and ice still would allow planetesimal-like objects to form, but the slower velocities and longer time between encounters would keep them from accreting to form full-sized planets. The lack of nearby planets to perturb the orbits would allow these objects to remain as remnants of the original solar nebula. The 30 or so Kuiper-belt objects that have been discovered so far are thought to represent the largest of a large population of icy planetesimals. There are estimated to be perhaps 70 000 objects larger than 100 km in size. Pluto may be just one of the largest of these objects. And Triton, a satellite of Neptune, may be a Kuiper-belt object that was captured into orbit rather than accreted.

We can see that there is a tremendous amount of regularity in our solar system, and that much of this regularity stems from the processes by which the planets were created. The inner planets, the gas giants, and the comets all seem to be a natural part of the collapse and evolution of giant molecular clouds.

Concluding comments

Are Earth-like planets common throughout the galaxy? As will be discussed in the next chapter, we are beginning to discover planets around other stars. Most of these newly found planets, however, are more massive than Jupiter. We do not have the ability yet to discover planets that are the size of Earth. How, then, can we extrapolate on the basis of our own solar system to understand the occurrence of planets elsewhere? We have developed a theory for how our solar system formed out of the remnants of the same collapsing giant molecular cloud that formed our Sun. This theory is rather simple and elegant, despite its apparent complexity. It relies on straightforward physical processes having occurred in ways that make sense to us intuitively. Although the details of the theory are still being debated, there is a general agreement on its basic building blocks.

This theory is able to explain the wide variety of objects that occur in our solar system; that is, it is consistent with the observed characteristics of our solar system. It also is consistent with what we see elsewhere, in particular with the occurrence of disks around a substantial fraction of the young stars that we can observe. These disks are thought to be protoplanetary disks and solar nebulas that may form into planets. In some observed disks, there is even a lack of debris close in to the star, suggesting that planets, although not observable, may already have formed.

Somewhere between one quarter and one half of the young stars that we can observe have disks surrounding them. If such disks do, in fact, evolve into planetary systems, then we expect that planets should be widespread. In addition, based on the physics of the accumulation process as seen in the computer simulations, we expect that a natural consequence of this should be the occurrence of planets very similar to our terrestrial planets. And, as a natural part of the accretion from disks, these planets should tend to form at about the same distance from their central star as the terrestrial planets are from the Sun. In other words, Earth-like planets should be widespread throughout the galaxy. The discussion in the next two chapters will center on the observational evidence for the existence of planets around other stars and on our expectations for their habitability – their ability to sustain liquid water and, therefore, life.

244   The detection and characterization of planets outside around other stars has long been one of the 'holy grails' of astronomy. Although theories have become well developed to describe the processes by which planets are thought to form (Chapter 14), those theories were developed to fit the observed properties of only a single planetary system – our own. We have known for some time that our Sun is a typical star in our galaxy. Since the early 1980s, we have known that flat disks of gas and dust, like that which is thought to have preceded the formation of our own solar system, exist around a large fraction of newly formed stars. Until we can determine whether other solar systems exist and whether they have properties like our own, however, there will be no way of knowing whether planetary systems – and Earth-like planets in particular – are common or rare in the galaxy. This will, of course, help us to understand whether life could be common or rare within the galaxy.

Several times during the past two decades, scientists have announced the discovery of a planet orbiting around another star. Until 1995, however, all of these discoveries either turned out to be erroneous or not repeatable. All of this changed in 1995. A number of groups of astronomers had been carrying out observational programs to search for planets for much of the previous decade. The techniques, to be described in detail below, were such that a single observation could not detect a planet. Rather, a series of very careful measurements, spread out over many years, were required in order to obtain convincing evidence. When the first true discovery of a planet-like object around another star came in 1995, it made front-page headlines in newspapers around the world. This was followed by a series of announcements from several different research groups with the confirmation of the existence of many different planets around different stars. By the end of 1996, close to a dozen true planets in other solar systems had been confirmed, along with a number of objects that are bigger than planets but smaller than stars. Another dozen or so have been suggested but not yet confirmed by additional observations. And, comments from the scientists involved suggest that this is just the tip of the iceberg – the existence of many more planets almost certainly will be announced during the coming months and years. One scientist has suggested that we are in the 'planet a month' phase of discovery.

What kinds of planets are being discovered? And, what do these tell us about the nature of planets and solar systems in general, the formation of planets, and the likelihood of Earth-like planets being out there?

The discovery of planets around other stars

In order to understand what these planets mean for the general occurrence of planets in the galaxy, we need to understand how they are detected and what can be learned about the planets from the observations. We will see that objects are being discovered that range from Jupiter-sized objects that are true planets up to small stars that can sustain nuclear reactions; each type of object can be detected by different means. It is important to realize that, even for the closest stars, a planet the size of Jupiter cannot be imaged directly – it is just too small and faint compared with the star that it orbits, and it would be lost in the glare from the star. Thus, most of the newly discovered objects have been detected by indirect means, based on their gravitational influence on the stars around which they orbit.

As a massive planet orbits around a star, each pulls gravitationally on the other, and both the planet and the star actually move around the center of mass of the system. This is analogous to a parent swinging a child in circles while playing 'airplane'. The child is much less massive than the parent and swings in a larger circle. Even the parent moves in a circular pattern, however, to keep from being pulled over by the looping child. If the parent and child were the same size, they would loop in circles that were the same size. Similarly, the planet in orbit around a star swings in a much larger orbit, but the pull of the planet on the star causes it to swing in orbit as well. In our own solar system, Jupiter is the most massive planet, at about 0.001 times the mass of the Sun; the center of gravity, around which both of them circle, is actually located just outside of the visible surface of the Sun, on the line connecting the center of the Sun and Jupiter (Figure 15.1).

The search for planets then becomes a search for the back-and-forth motion of the star as a planet orbits around it. This motion is known as the 'reflex' motion of the star. There are two easy ways to try to detect this motion, and these methods are responsible for most of the known planetary objects. First, it is possible to detect the changes in the velocity of the star relative to the Earth; as the planet moves around the star, the star will move first toward us and then away from us. This is the 'Doppler' technique, because it uses the changes in the wavelengths of light emitted from a star that result from its having a velocity either toward or away from us (the so-called 'Doppler shift'). Second, careful observations of the location of a star against the background sky can detect this wobbling reflex motion directly. This technique is called the 'astrometric' technique, based on astrometry, the careful measurement of the direction to astronomical objects.

[Figure 15.1]
Schematic diagram of the center
of mass of the Jupiter–Sun system
(not to scale). Jupiter's mass is 1.05
$\times 10^{-3}$ times that of the Sun, so the
center of mass is located about
1000 times closer to the center of
the Sun than to Jupiter. The center
of mass is 8.17×10^5 km from the
center of the Sun, just outside of
the visible surface of the Sun at
6.96×10^5 km from the center.

In each case, one can imagine that the motion of the star will be larger for a more-massive planet. Smaller planets will produce reflex motion in a star as well, but the motion will have a smaller amplitude. Thus, we expect that the first detections of planets will be of the most massive planets. Also, it is clear that, to see the full back-and-forth motion of a star that would provide convincing evidence of a planet's existence, one would need to observe the velocity of a star many times during the time that it takes the planet to orbit once around the star. Jupiter, for instance, goes around the Sun once every 12 years, and the detection of a Jupiter-like planet around another star would require observations spanning a decade or longer. This will favor the detection of planets that orbit closer in to their central star and have the shortest orbital periods.

It is worth a brief digression to discuss the classification of objects of different masses – what is the boundary between stars and planets? Multiple stars are quite common; in fact, about 2/3 of the nearby stars are in binary (two-star) or multiple star systems. In these systems, two or more stars orbit around each other. Binaries and multiples form when a collapsing giant molecular cloud breaks apart into separate clouds as it collapses. This breakup can be triggered by the initial rotation of the cloud; the collapsing cloud spins fast enough that it tears itself apart into multiple collapsing clouds. Each cloud fragment will continue its collapse separately and will become a separate star. Often, the breakup will produce clouds that are rotating around each other; the stars that result, then, end up in orbits around each other.

The most massive stars known are about 20–50 times the mass of our Sun (20–50 M_S), and our Sun seems rather typical of stars. The most massive stars, though, have higher interior pressures and temperatures that trigger a more-rapid fusion of hydrogen atoms into helium; it is this nuclear fusion process that releases energy in the star. The rapid burning of the nuclear fuel in the larger stars results in their having very short lifetimes, in some cases as short as 10 million years. At the other extreme, stars as small as only 8% of the mass of the Sun (0.08 M_S) still have a high enough central pressure and temperature that hydrogen fusion can result. Objects smaller than this cannot sustain nuclear reactions, so this is the smallest that an object can be and still be considered a star. A number of these smallest stars have been detected orbiting around other, more-massive stars, using the same techniques that are used to look for planets.

During collapse of a giant molecular cloud and its breakup into smaller clouds, the smallest cloud that can collapse into an isolated object is thought to be about 1% the mass of the Sun. These objects are too small to burn nuclear fuel, but still give off energy. Their energy comes from the conversion of gravitational potential energy to heat as they collapse. During their youth, they can give off more light than our Sun does at present. Those objects that form by the collapse of molecular clouds, but are too small to trigger the fusion of hydrogen, are referred to as 'brown dwarfs'. The phrase 'dwarf' is used to indicate that they are less massive than the usual stars, and 'brown' means that they give off only meager amounts of light over their lifetimes (one step above black, giving off no light). Thus, objects that are more massive than 8% of the Sun's mass generally are considered to be stars, and those between 1 and 8% are brown dwarfs.

Jupiter is less massive than the smallest possible brown dwarfs by another factor of ten. Thus, brown dwarfs span the range between about 10 and 80 Jupiter masses (M_J). Of course, Jupiter is also thought to have formed by a different process than brown dwarfs, involving the accumulation of a rocky core in a protoplanetary disk and the subsequent attraction of gas from the disk (see Chapter 14). Theories of the formation of the giant planets suggest that the largest planet able to form by this process would have a mass not much more than that of Jupiter.

Thus, on this theoretical basis, we expect that there might be two different classes of substellar objects (smaller than stars): those that formed from a disk and cannot be much more than a few times as massive as Jupiter – these would be called planets; and those that are between about 10 and 80 M_J in size and formed from the collapse of molecular clouds – these would be called brown dwarfs. Although there is a school of thought that says that it does not matter what we call these objects – that they all are part of a continuum of sizes from planets up through stars – we can distinguish the objects on the basis of their mechanism of formation; this is the approach that we have taken here. Of course, observations of an object orbiting around a star do not allow a direct determination of the process by which it formed; thus, the distinction between planets and brown dwarfs can be based only on theory. Regardless of their mechanism of origin, planets, brown dwarfs, and the smallest stars together make up the category of objects called stellar companions – they are objects orbiting around a star.

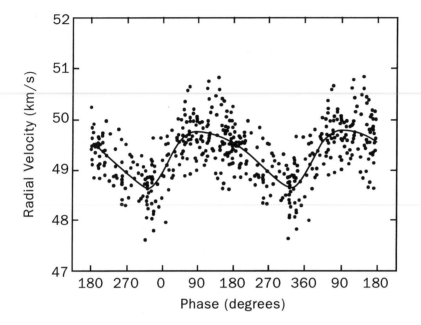

[Figure 15.2]
Radial velocity measurements used to demonstrate the existence of a brown dwarf around the star HD114762. Dots represent the individual measurements, and the smooth line is the value expected based on the best-fitting model of an orbiting brown dwarf. (After Cochran *et al.*, 1991.)

The first candidate brown dwarf was detected in 1989 using the 'Doppler' technique. The object was found orbiting the star HD114762 (the name comes from its description in one of the many catalogs that have been compiled to list and describe various astronomical objects). This star is thought to be very similar to our Sun, having a similar mass and surface temperature, although with a lower abundance of the elements heavier than helium. The radial velocity of the star was measured by looking at the shift in the wavelength of certain key absorption features seen in the star's atmosphere; the wavelength shift is determined by the velocity, and the direction of the shift (toward shorter or longer wavelengths) by whether the velocity is toward or away from us. The velocity of HD114762 varies by about 500 m/s, first toward and then away from us, with a period of about 84 days (Figure 15.2).

The mass of the companion and the size of its orbit around the star can be determined from the observations, as follows. The central star's mass can be determined from its temperature and from the total amount of light that it emits. This is done using the 'Hertzsprung–Russell' diagram (named for the astronomers who first developed it), which is compiled from these properties for various nearby stars. From the mass of the star and the period of the orbit of the companion object, the distance of the companion from the star (the size of its orbit) can be determined. And, given the mass of the star and the amplitude of the reflex motion, derived from the observed variation of the star's velocity with time, the mass of the companion can be determined. From the details of the variation of the star's velocity with time, the eccentricity of the orbit – whether it is more circular or more elongated – can even be determined.

There is a major complication in this analysis, however. The Doppler-shift technique is a way to determine the velocity of the star toward or away from the observer, along the 'line of sight' between the object and the Earth. If the velocity is not directly toward or away from the observer, then only that component of the velocity that is along the line of sight can be determined. For example, if the companion's orbital plane around the star is aligned with this line of sight, then the maximum measured velocity equals the maximum actual velocity. However, if the plane is misaligned, then the actual velocity will be larger than the measured velocity. At the extreme, if the plane of the orbit coincided exactly with the plane of the sky (in other words, if we were looking down on the solar system from above), then there would be no velocity toward or away from us as the companion orbited the star; this would be the case regardless of the actual velocity. And, if we were looking at the orbital plane at an oblique angle, then the component of the velocity toward and away from us would be only a fraction of the actual velocity along the orbit. For this reason, the mass that is determined for the companion using the Doppler technique is only a lower limit on the actual mass, and the actual mass can be greater depending on the orientation of the companion's orbit. A different way of saying it is that the derived mass of the companion is really given by the product of the true mass and the sine of the angle measured between the line of sight to the Earth and the line normal to the orbital plane of the companion (parallel to the pole of the orbit).

For HD114762, this lower limit on the mass is about 11 M_J (0.011 M_S), at the lower end of the size range expected for brown dwarfs. For this object, however, there is an additional observation that helps to determine the orientation of the orbit. The average Doppler shift measured for the stellar atmosphere absorption line determines the overall velocity of the star. The shape of the absorption line (how it varies with wavelength), however, is determined by the emission from the different visible parts of the star. The overall shape includes the Doppler shift produced as the star rotates, with part of the star moving toward us and part away, and from turbulence within the star itself. Comparison with similar stars allows an estimate to be made of the contribution of turbulence, so that the remaining parts of the shape of the line tell us about the rotational velocity of the star. Because we know the period of rotation of the star, from changes in its output

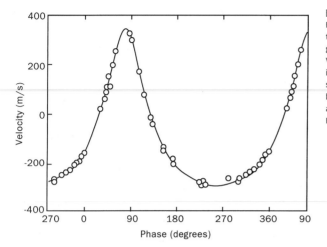

[Figure 15.3]
Radial velocity measurements used to demonstrate the existence of a gas-giant planet around the star 70 Virginis. The circles represent the individual measurements, and the smooth line is the value predicted based on the best-fitting model of an orbiting planet. (After Marcy and Butler, 1996.)

similar to those we see in the Sun due to sunspots, we can infer the orientation of the stellar polar axis and, most likely, of the companion's orbital plane. In this case, we seem to be seeing the orbital plane nearly face on (that is, from above), and the actual mass of the companion is inferred to be greater than about 20 M_J (0.02 M_S) and possibly as great as 100 M_J (0.1 M_S). Thus, the substellar companion to HD114762 is on the boundary between being a brown dwarf and being a small star.

In 1995, three additional substellar objects were discovered using the Doppler technique. These were companions to the stars 51 Pegasi (in the constellation of Pegasus), 47 Ursae Majoris (in Ursa Major), and 70 Virginis (in Virgo) (Figure 15.3). The properties of these objects span a remarkably wide range. The masses (lower limits, again, because of the uncertainty in the orbital orientation) are 0.5, 2.4, and 6.6 M_J, respectively. Statistically, we can estimate the odds of the orbital plane having different orientations, and thereby guess at the actual mass. If the orientation of the companion orbits is completely random in the sky, there is a 2/3 chance that the actual companion mass is less than about 1.5 times these values, and only a 5% chance that the actual mass is greater than about 3 times these values. Thus, the first two objects probably classify as planets, and the third may be a brown dwarf.

The distance at which these planets orbit around their central stars is about 0.05, 2.1, and 0.45 A.U., respectively. These values are striking in comparison with our own solar system, because these planets' orbits are much smaller than Jupiter's orbit. In addition, the first two planets are in nearly circular orbits, while that around 70 Virginis is in a very elliptical orbit. Clearly, these solar systems are very different from our own, and something very intriguing is going on here. If we are to understand the behavior of other solar systems, we must learn how these very different planets evolved.

The astrometric technique of looking for planets has resulted in the discovery of a single planet so far, orbiting the star Lalande 21185. The direct measurement of the reflex motion of the star in two dimensions (on the background sky) as the planet moves around it allows the ambiguity of the orientation of the plane of the orbit to be removed, so that the actual mass of the planet can be determined. In addition, with Lalande 21185, the motion is more complicated than just a single back-and-forth motion; rather, it is two different motions with different periods, added together, suggesting that there are two different planets orbiting the star. The best fit to the data indicates that these two planets have masses of about 1.0 and 1.5 M_J, and that they are orbiting at distances of 2.5 and 10 A.U. from the star, respectively. The larger of these planets is, in fact, very much like Jupiter in our solar system. Some caution must be used in accepting these particular planets, however; the orbital period of the larger planet is about 30 years, so its presence can be confirmed only when we have taken measurements over a period of nearly 30 years.

Although imaging generally is not thought to be a useful tool for searching for extrasolar planets, it has been used to discover one substellar companion. The technique involves using what is called 'adaptive optics' on a telescope to improve the quality of a telescopic image, combined with a mask that can absorb most of the light from the central star. This technique was used on a number of nearby stars, and allowed the discovery of a companion to the star Gliese 229 (GL229) (Figure 15.4). The companion (GL229B, with the B indicating the companion) both reflects starlight and emits its own light from the heat left over from its formation and from its contraction. From the characteristics of the emitted light, its surface temperature is inferred to be less than 1000 K. This is much cooler than would be expected for even the smallest star burning nuclear fuel. In addition, the spectrum shows that carbon is present as CH_4 rather than CO, also consistent with this low temperature. As a result, the companion is thought to be a brown dwarf. Interestingly, the emission spectrum of GL229B looks remarkably similar to that of Jupiter, suggesting that it has a very similar composition (Figure 15.5). From the total amount of emitted energy, the mass of Gl229B is thought to be between about 20 and 50 M_J, again well within the range expected for a brown dwarf.

Finally, there is one more planet category that has been discovered – planets around pulsar stars. A pulsar consists of the remains of a star that

252

[Figure 15.4]
Hubble Space Telescope image of the brown dwarf Gliese 229B. GL229B is the bright spot near the center of the image. The central star, GL229, is overexposed in the upper left, and the spike extending toward the lower right is an instrumental artifact. (NASA photo.)

has undergone a supernova. Pulsars consist of rapidly rotating objects about the size of the Earth and composed entirely of neutrons – neutron stars. The rapid rotation induces the emission of radio waves, and the rotation causes the beam of radio emission to sweep around the sky, much like a searchlight on top of a lighthouse. When the beam passes through the Earth, we detect it as a pulse of radio emission. These objects rotate at a nearly constant rate, so the time between pulses is nearly constant. For two of these pulsars, however – PSR B1257+12 and PSR B0329+54B (here, the name is used to give the approximate location of the object in the sky) – very regular changes are seen in the timing of when the pulses are received. These changes are interpreted as resulting from the back-and-forth reflex motion of the neutron stars as planets orbit around them. The change in pulse timing results from the change in the distance that the radio beam has to travel from the neutron star to us as it moves back and forth. The first of these pulsars shows pulse timing changes with three different periods

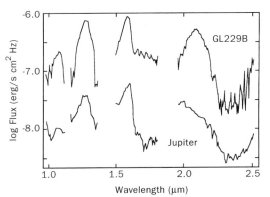

[Figure 15.5]
Emission spectrum of the brown dwarf GL229B, compared to that of Jupiter. The similar nature of the features are expected for a brown dwarf. (After Oppenheimer *et al.*, 1995.)

superimposed on each other; this suggests the presence of three planets, having minimum masses of about 2.8, 3.4, and 0.015 times the mass of the Earth (M_E). The second has a single planet with a minimum mass of about 2 M_E. Planets the mass of the Earth can be detected in this case because of the remarkably regular period of the radio emissions; even the very small changes in the period that result from this very small reflex motion can be detected. Orbiting around a pulsar is an odd environment. No planet, let alone life, could survive the supernova formation of a pulsar; these planets must have formed from leftover debris following the pulsar formation. These results, however, provide convincing evidence for the existence of Earth-sized (if not Earth-like) planets!

The diverse 'zoo' of planets and brown dwarfs that have been found is summarized in Figure 15.6.

How to explain the 'oddball' planets

The list of confirmed substellar companions presents quite an odd group. There are four planets orbiting around pulsars, all approximately Earth or Moon-sized. The rest are Jupiter- sized or larger. A number of them are in the mass range of 1–10 M_J or greater. One of them clearly is a brown dwarf (GL229B), being between about 20 and 50 M_J. And one (HD114762) could be a Jupiter-sized object or, given the high apparent inclination of its orbit to the line of sight to the Earth, might be a small star. The abundance of Jupiter-mass and larger objects is not a surprise – we expected to be able to detect these most easily and, based on our own solar system, we expect to find gas giant planets around other stars as well. However, many of the

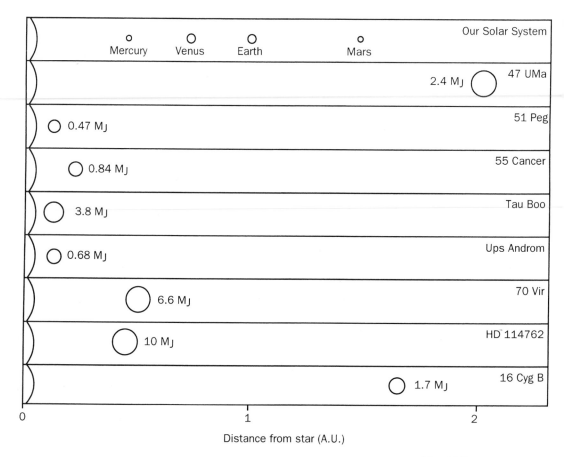

[Figure 15.6]
Schematic drawings of the eight
confirmed solar systems around
other stars, compared to the
planets in our own inner solar
system. In each case, the planets
are shown at their appropriate
distance from their central star.
Each extrasolar planet is labeled
with the star name and with the
most probable value of its mass.
(After Marcy and Butler, 1997.)

Jupiter-sized objects have orbits that are very close in to their star, deep
within the part of the solar system where we expected smaller, rocky
planets to exist. In order to understand what the implications of these
planets are for the possible existence of rocky planets, we need to consider
the formation and evolution of gas-giant planets in more detail.

Although we discussed the formation of the terrestrial and the gas-
giant planets in our solar system in Chapter 14, we did not address specifically what might have caused the giant planets to form where they did. In
our inner solar system, the amount of material with rocky composition
that was present in the protoplanetary disk is thought to be just about the
amount required to form the terrestrial planets. The accretion into rocky
bodies occurred rapidly, probably well before the dissipation of the
remaining gas, but the final planets were not massive enough to capture
and hold any nebular gas.

We infer that the rocky cores that formed in the vicinity of Jupiter and Saturn were large enough to hold onto gas in the disk. Otherwise, Jupiter and Saturn would not have thick atmospheres that have the same composition as the Sun. Calculations suggest that the rocky cores would have to be about 10–20 times the mass of the Earth in order to begin to hold onto hydrogen and helium. Are these results consistent with our theories of how the planets formed? Our models of the formation and evolution of disks suggest that they are densest closer in to the star and less dense farther out; farther out, though, the disks have more surface area. The product of the density of the disk and the surface area at a given distance from the Sun gives the total mass available to form planets at that distance. Unfortunately, the estimated distribution of the rocky material in the disk results in insufficient mass in the vicinity of Jupiter and Saturn to form the massive rocky cores that are required. Something like 10–100 times as much mass must be present in the disk in order to allow the rocky core to form fast enough for gas to be accreted from the nebula. Where does the rest of the mass for the rocky core come from?

One possibility is that the rest of the mass consists of water ice. The temperatures in the protoplanetary disk were lower farther out from the proto-Sun. This is because of the greater distance away from the proto-Sun's source of heat and because there was less compressional heating of the disk during its collapse at that distance. Although the temperature at which water vapor will begin to condense into ice is somewhat uncertain (since it depends on the pressure of the gas in the nebula), it cannot be too different from about 160 K. Once water ice is able to condense, the formation of rock-and-ice planetesimals becomes possible. The tremendous availability of hydrogen and oxygen in the solar nebula means there might be as much as ten times more water ice than rocky material. The central core of the gas giants, therefore, may have formed from a combination of icy and rocky material. In this way, there would have been enough mass available to build rapidly the massive cores that can attract and hold on to the gas from the disk.

This approach suggests that the formation of the gas giants depended on the condensation of water ice. If this theory is correct, then it provides a simple explanation for the compositional variations of the planets in our solar system (as described at the beginning of Chapter 14). Where only rocky material was available, close in to the Sun, the rocky planets formed – the four terrestrial planets Mercury, Venus, Earth, and Mars. Where water ice also was available to accrete, massive cores could form quickly enough that they would be able to hold onto the hydrogen and helium in the nebula before it dissipated – the gas-giant planets Jupiter, Saturn, Uranus,

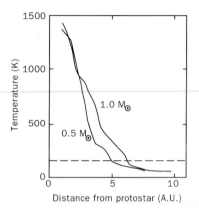

[Figure 15.7]
Temperature at the mid-plane of
protoplanetary disks as predicted
by a model of their growth and
evolution. Models are shown for
two different disk masses, 1.0 and
0.5 times the mass of the Sun.
Temperatures drop to about 160 K,
low enough for water to begin to
condense, at a distance of 5–7 A.U.;
this is very close to Jupiter's
distance from the Sun, and the
condensation of water may be
responsible for allowing Jupiter to
have formed. (After Boss, 1995.)

and Neptune could form. The location of the gas giants, then, would depend on the location of the water-condensation line. Gas giants could form only farther out than the distance at which water ice could first condense. Models of the protoplanetary disk that formed our solar system suggest that water condensation probably occurred very close to the present orbit of Jupiter (Figure 15.7).

If we extrapolate this result to other solar systems, we expect that gas giants will form only in the outer parts of protoplanetary disks, out past the distance where water ice can begin to condense. Models of the formation and evolution of protoplanetary disks suggest that this distance should not be too different for disks of various sizes or around protostars of various masses. It clearly should not be possible for gas giants to form closer than several A.U. to their central star. The presence of so many oddball planets, with Jupiter-like planets so close to their star (closer than Mercury is to our Sun), says that something is wrong with this picture. Either these planets formed in a protoplanetary disk that was 10–100 times denser than ours, or they formed in the outer parts of their solar systems and their orbits evolved inward. The first possibility is unlikely, as that would require that the protoplanetary disks contain too large a fraction of the total mass available in the system; that leaves the second possibility.

If we imagine that these planets formed farther out in their solar systems, then how could they have evolved to their present locations? One possibility involves gravitational interactions between two gas-giant planets in the outer solar system. If two gas-giant planets formed near each other, each would exert a gravitational pull on the other. If they are close enough together, this pull will be strong enough to perturb the orbits and, eventually, the two planets may have a close encounter with each other. As a result of this encounter, the two objects exchange momentum; one planet can be flung into an orbit that will send it out of the solar system, **257** and the other could be swung into an extremely elliptical orbit that would bring it very close in to the star. This process is similar to the gravitational slingshot effect that was used by the *Voyager* spacecraft; it was sent close to Jupiter in order to fling it out toward Saturn. In this instance, however, the spacecraft was so small that the effect on Jupiter's orbit was negligible. Once one of the gas giants was in an elliptical orbit, tidal interactions with the star itself could have circularized the orbit. This would leave it close to the star in a final, stable orbit similar to some that have been seen.

This theory may not be as implausible as it sounds. As gas-giant planets form, they are thought to wander outward as a result of the gravitational scattering of a large number of planetesimals out of the solar system. This wandering can bring two planets that formed a substantial distance away from each other into closer proximity, where they can begin to interact with each other gravitationally.

The second possible mechanism for moving a giant planet involves the dynamical interactions between the newly formed planet and any remaining disk material. These interactions make it seem as if the planet is moving through a viscous disk, and they take energy from the planet's orbit and cause the planet to spiral in toward the star. The interactions take the form of what are called 'spiral density waves'. Imagine a planet pulling on a dust grain that is in an orbit just inside the orbit of the planet. The dust grain moves faster through its orbit, so will approach the planet from behind, catch up with it, and then pass it. As the dust grain is catching up with the planet from behind, the planet pulls it in the forward direction in its orbit, and it speeds up a little bit. After the dust grain passes the planet, the planet's gravity pulls it in the backward direction and it slows down again. If nothing else happens, then this forward pull and backward pull are equal and opposite, and there is no net change in the velocity of the grain. In a thick disk, however, the particles are constantly colliding with one another. This means that before the planet can tug in the backward direction and balance out the forward pull, the impulse is transferred by collision to other particles. This produces a net pull on the dust grains and a net transfer of momentum from the planet to the dust grains. As a result, 'waves' of motion are set up in the dust grains, they carry momentum and energy away from the planet, and the planet will spiral toward the star. This process can be quite effective in moving the planet from its original location toward the star.

These spiral density waves have been seen, for example, in Saturn's rings (Figure 15.8). There, small satellites play the role of the perturbing object, and ring particles play the role of the perturbed debris. Spiral density waves also play a role at the galactic scale, since they are thought to

[Figure 15.8]
Detailed structure in Saturn's rings. The image was constructed from a brightness trace across the rings obtained from the *Voyager* spacecraft, and shows fine-scale detail in the rings. This structure results from the generation of spiral density waves. (Image courtesy L. Esposito.)

be responsible for the presence of the spiral arms in galaxies like our own.

With this mechanism of a planet spiraling in toward the star, the only problem is how to stop it from coming in all the way and colliding with the star. Stellar tides may slow and stop the inward spiral, or the spiral may stop when the planet enters a region close to the star where disk material has been removed by interactions with the star's magnetosphere.

Although we clearly do not understand fully how planets can evolve inward in their solar systems, it is clear that possible mechanisms do exist. It is easier to imagine that these oddball giant planets formed far out in their solar systems and evolved inward than it is to imagine a process by which they could have formed anywhere near to their present locations.

Are extrasolar planets common?

What kinds of planets and solar systems are typical? Do most gas giants form in the outer parts of solar systems and stay there, or do most of their orbits evolve inward toward their central star? This is not a trivial question, as the answer will determine if Earth-like planets are likely to be present around other stars. If the giant planets have evolved in toward their central star, by either of the above mechanisms, then they very likely would have had a major effect on any rocky planets that were present. If the gas giant was slung into an elliptical orbit, it probably would have ejected any rocky planets from their solar system or pulled them in to accrete onto the gas giant. If the gas giant spiraled in toward the star gradually, it almost cer-tainly·would have accreted much of the mass that it encountered on its way. In either instance, the possibility of an Earth-like planet being present would be decreased substantially.

Jupiter-sized planets seem to be distributed throughout their solar systems, from very close to their central star to very far away. At present, we have not discovered enough planets to try to guess which behavior would be more typical, or even what the range of possible behaviors might be.

However, we can begin to place limits as to what fraction of stars out there actually have planetary systems surrounding them. First, we note that between about one quarter and one half of all very young stars appear to have disks similar to the protoplanetary disk out of which our solar system is thought to have formed (see Chapter 14). This would seem to be an upper limit as to the fraction of stars that have planets – if no disk formed originally, there can be no planets today.

Second, the searches that have been conducted for planets have found a small number of planets after searching a large number of stars. Somewhere between about 3 and 6% of the stars examined have been found to have planets. However, planets are continuing to be discovered around these same stars as more data are obtained. Some stars may have gas-giant planets that are too small to be discovered with the currently available technology. And, some stars may have Earth-like, rocky planets but no gas-giant planets. Thus, these estimates would seem to be a lower limit on what fraction of stars might have planets.

It should be clear that, despite the tremendous excitement of discover-ing planets around other stars, and despite the major significance of these discoveries for furthering our understanding of the distribution of planets in our galaxy, there still is insufficient information to help us determine how many planets are out there. We are still stuck with the theories of planet formation as the major driver of our understanding of the distribu-tion of planets in our galaxy. Based on the available statistics, however, we can conclude that planets in general, and Earth-like rocky planets in par-ticular, probably are very common throughout the galaxy.

260 The possibility of life existing on planets around other stars centers not only on the existence of planets there (Chapters 14 and 15), but also on whether environmental conditions on those planets permit life to exist. The basic requirements for life are thought to be the occurrence of liquid water, access to the biogenic elements that are the building blocks of life, and a source of energy able to drive chemical reactions resulting in the production of molecules that are out of equilibrium with their environment (see Chapter 7). In our own solar system, these requirements are met with certainty today only on the Earth. They have been met on the Mars surface and subsurface at various times during the last 4.5 billion years, and conceivably could be met within a subsurface ocean on Europa. What is the likelihood of these conditions being met on planets that might exist around other stars? Using information from our own solar system regarding how planetary climates and environments work, we can try to address this question.

Other stars appear to be made of the same elements that the Sun is made of. This is not surprising given that the heavier elements were created within preexisting stars and distributed throughout the galaxy. We imagine, then, that most extrasolar rocky planets will be made from the same elements of which the Earth is made, although possibly in different proportions. This means that there should be access to the important biogenic elements either at or near their surfaces. And, the existence of a source of energy is relatively straightforward, be it from starlight or from geothermal sources. As in our own solar system, the availability of liquid water becomes the most difficult requirement to satisfy. What aspects of a planet's atmosphere control the climate and determine the availability of water? What are the factors that will make a planet too cold or too hot to allow liquid water and, therefore, life to exist? The answer to this last question may seem trivial – if a planet is too close to a star it will be too hot; if it is too far away, it will be too cold. Because of the complex nature of climate, however, the details can become quite complicated. There is not a simple answer as to what makes a planet too cold or too hot, although there are some simple approaches that will give us much insight.

The habitable zone around a star (HZ) is defined as the region where a planet could reside and maintain surface temperatures that allow liquid water to exist. Outside of this zone, temperatures will be either too cold or too hot to allow liquid water to exist. The HZ will be defined at a given instant in time, based in part on how much energy the star emits. The energy output of a star varies with time, however, so that the location of the habitable zone also will vary. As the energy given off by a star increases throughout its lifetime, the HZ will be pushed outward. Thus, in order to study the evolution of a planet's climate through time, and the possibility **261** of life existing for a long period of time, we need to look at the history of the HZ. We can define the continuously habitable zone (CHZ) as that region around a star that is located within the habitable zone for a period long enough for life to exist. For our own solar system, the CHZ would be that region where liquid water could have existed for the duration of the 4.5 b.y. of Earth's history.

Habitability of planets in our solar system

Our own solar system contains four terrestrial planets within the inner part of the solar system. Three of these planets provide substantial clues as to what might determine the boundaries of the habitable zone. We will discuss each planet's climate briefly, and then look in more detail at the processes that determine the location of the boundaries of the HZ.

The Earth, of course, is habitable. Abundant liquid water is present at its surface, with 97% of the water in the oceans, 2% as permanent ice cover in Greenland, Antarctica, and the Arctic Ocean, and the rest as lakes, streams, and groundwater. A small amount of water is present in the atmosphere as water vapor that can condense to form clouds, rain, and snow. Life exists in a wide variety of locations near the Earth's surface, essentially wherever liquid water is present (see Chapter 5). These include the surfaces of the continents, where the average temperature is about 288 K (15 °C, or 59 °F). In the deep oceans, life exists in water that is very close to the freezing point of ice (273 K, 0 °C, or 32 °F). In some locations where the presence of salts has lowered the freezing point, it is possible for life to exist at temperatures just below the usual freezing point. Life can survive short periods at below-freezing temperatures (such as during the deep freeze of a New England winter), as long as the temperature rises above melting and allows liquid water to exist during parts of the year. At the other extreme, we know of life that exists at temperatures of up to about 115 °C (239 °F), in hot springs, hydrothermal vents, and geothermal areas. Terrestrial life is thought to be able to exist at temperatures of up to about 150 °C (about 300 °F).

The climate of the Earth is controlled by the competing effects of the ability of sunlight and infrared thermal energy to be transferred through the atmosphere, the evaporation and precipitation of water, and the climatic effects of living organisms. An important part of the climate system is the role of the atmospheric greenhouse gases, primarily CO_2 and H_2O.

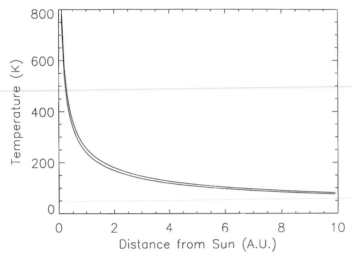

[Figure 16.1]
Equilibrium surface temperature as a function of distance from the Sun. The top curve shows values today, and the bottom curve shows values 4.5 b.y.a., when the Sun was 30 % dimmer. Both curves are for rapidly rotating planets without any atmosphere.

Without them, the Earth's temperature would be about 255 K, well below the freezing point of water, and the Earth would be outside of the habitable zone around the Sun (Figure 16.1). Although we often think of humans as affecting the climate, through their role in the emission of CO_2 into the atmosphere within the last century, other organisms have been affecting the climate for billions of years. For instance, the presence of abundant oxygen in the atmosphere, beginning about 2 b.y.a., resulted from the metabolism of single-celled organisms.

Venus, on the other hand, has a thick, CO_2 atmosphere that drives a greenhouse warming that heats the surface up to a temperature of about 750 K (Chapter 11). There is very little water vapor in the atmosphere, and temperatures are too hot for water to be present on the surface as either liquid water or water of hydration in minerals. The total amount of CO_2 on Venus, however, is about equal to that on the Earth. While most of the Earth's CO_2 is tied up in carbonate rocks (such as limestone and marble), most or all of Venus's CO_2 is in the atmosphere. Venus appears to have suffered a catastrophic response to the larger amount of heating that results from its being closer to the Sun. The explanation for the high temperature is the occurrence of a runaway greenhouse (see Chapter 11). The greater solar heating resulted in the evaporation of a large amount of water from an early ocean into the atmosphere. Since water vapor is a greenhouse gas, the increased water abundance led to higher temperatures, which led to even more water evaporation, and so on. Ultimately, all of the water from an ocean could have been in the atmosphere along with all of the CO_2, and surface temperatures would have become extremely high. The water near the top of the atmosphere would have been broken apart into oxygen and hydrogen by ultraviolet light from the Sun, and the hydrogen could have escaped to space. Once the CO_2 was in the atmosphere, it could not be easily removed again. This would have left the CO_2 behind in the warm, stable atmosphere that we see today. The situation is more complex than this, however, because of the role played by volcanic SO_2 gas in the total greenhouse warming.

Clearly, Venus is inside of the inner boundary to the habitable zone around the Sun, at least for a planet of its size and atmospheric composition. It is possible, though, that when the energy emitted from the Sun was lower during the early history of the solar system, Venus might have been outside of the inner boundary of the HZ and had a more clement climate.

Finally, consider Mars (see Chapter 8). Mars shows abundant evidence for the occurrence of liquid water at its surface throughout time, even though liquid water is not stable at the surface today (it would either freeze or evaporate quickly). The first half billion years of history recorded in the martian geology (from about 4.0 to 3.5 b.y.a.), however, suggests that liquid water was much more stable then than it is today. There are tributary systems that look as if they were carved by water or water-rich debris flowing over the surface, and there is evidence for widespread erosion on the surface that probably was caused by liquid water. Since that time, however, liquid water probably has been present only on an occasional, transient basis. The catastrophic release of water from beneath the surface, in events that formed the outflow channels, occurred only intermittently, and any standing bodies of water such as lakes or oceans, if present at all, could have existed only for short periods of time.

Mars is outside of the habitable zone today (at least, as defined above), but appears to have been within it during its early history. It is quite puzzling in this regard: the climate appears to have been at its warmest, allowing liquid water, during the early epochs when the solar output was the least. As the amount of heating from sunlight increased during the last few billion years, the martian geological record suggests that the temperatures actually went down!

We seem to have the situation where all three planets allowed liquid water to exist on their surfaces during the earliest history of the solar system. As the Sun heated up over the 4.5 billion years since formation, the planet closer to the Sun heated up and lost its ability to retain liquid water, and the planet farther from the Sun cooled off and lost its ability to retain liquid water. Only the planet in the middle has been able to retain liquid water at its surface for most or all of that time period. Clearly, habitability of a planet is not simply a function of distance from the Sun. The behavior of a planet's atmosphere must be playing a role, and the climate clearly responds in some complicated way to the Sun's output. Thus, although we can define a habitable zone, this does not mean that any planet within the zone automatically will be habitable; it still must have an environment that can sustain liquid water.

There are two approaches that we can take, then, in our effort to understand the habitability of extrasolar planets. First, we need to explore the planet's climatic response to heating by the absorption of starlight; in essence, this determines the planet's surface temperature and its ability to allow liquid water to exist. Second, we must understand those intrinsic characteristics of a planet that allow it to maintain an atmosphere in the first place.

The terrestrial planets as a guide to planetary habitability

All three of these terrestrial planets – Venus, Earth, and Mars – have an atmosphere that has a warming due to the presence of greenhouse gases. In each case, CO_2 is the primary greenhouse gas, with H_2O and ozone (O_3) contributing on the Earth and SO_2 on Venus. Venus, with the most atmospheric CO_2, shows the largest greenhouse warming and Mars, with less CO_2 and less atmospheric water vapor, shows the smallest. Because of the relatively high cosmic abundances of carbon and oxygen, we expect that CO_2 will play a major role in the behavior of atmospheres in general. The fact that the martian atmosphere would be warm enough to support liquid water today if only it had more CO_2 in its atmosphere, and that the Earth would not be warm enough without any CO_2, tells us how important the CO_2 abundance and atmospheric history are for planet habitability. Similarly, many of the differences between the Earth and Venus can be explained by the fact that most or all of Venus' CO_2 is in its atmosphere while the Earth has little atmospheric CO_2 and the bulk of its carbon is contained in surface minerals. Which properties of a planet determine where the CO_2 will reside?

We can understand the Earth's CO_2 inventory and atmospheric abundance by looking at the ability of carbon to cycle between the atmosphere and the crust (see Chapter 4 and Figure 4.6). With liquid water present at the Earth's surface, atmospheric CO_2 can dissolve in the water, where it is removed by chemical interactions with surface rocks that form carbonate minerals. In particular, CO_2 can combine with calcium or magnesium atoms that are brought to the oceans by the runoff of water from the continents. Although the formation of carbonates on Earth is done predominantly by living organisms (marine animals that form shells out of the CO_2 dissolved in the oceans), many of the same minerals would form by strictly geochemical means even in the absence of life. The newly formed carbonate minerals sit on the ocean floor. There, the process of plate tectonics will slowly move the ocean floor toward subduction zones, like a conveyor belt, where it can be pushed beneath the surface of the continents and into the upper mantle. As the carbonates are subducted, the resultant heating at depth causes the CO_2 to be released from the minerals; the CO_2 becomes part of the volcanic magma associated with subduction zones, rises up to the surface, and is released back into the atmosphere. This recycling of CO_2 as a result of global plate tectonic activity serves to keep the atmosphere supplied with CO_2; without this resupply, all of the atmospheric CO_2 would be removed to form carbonates after only 400 000 years. There may even be an effective feedback mechanism between the atmospheric CO_2 abundance, the temperature, and the rate of carbonate formation (Chapter 3); this mechanism would serve to leave just enough CO_2 in the atmosphere to stabilize the surface temperatures and keep the oceans in liquid form.

Although Venus does not exhibit plate tectonic processes that recycle CO_2 back into the atmosphere, the high surface temperature has the same net effect. Any carbonate minerals that do form will be buried quickly (on a geological timescale) by volcanic material or will be squeezed beneath the surface at one of the small-scale subduction zones. At a relatively shallow depth, the higher temperatures will break apart the minerals and release the CO_2 back into the atmosphere. The steady-state equilibrium between the formation of carbonate minerals from the CO_2 in the atmosphere and the destruction at high temperatures acts to keep most, if not all, of the CO_2 in the atmosphere.

Mars, at about a tenth of the mass of both Earth and Venus, does not have any plate tectonics that can help to recycle carbonates. If there was liquid water present on early Mars (as inferred from the geology), and if an early, thick CO_2 greenhouse atmosphere was responsible for raising the temperature enough to allow water to exist, then we would expect that the formation of carbonates would be a natural consequence. Any carbonates that formed, however, would have been difficult to recycle back into the atmosphere. Although the geological observations suggest that warmer temperatures were present for perhaps as long as 500 million years, estimates of weathering rates suggest that the CO_2 could have been removed in less than 100 m.y. Some limited recycling could have occurred as a result of the volcanism that occurred throughout time – molten rock coming into contact with carbonate minerals would have heated the minerals and released the CO_2 back into the atmosphere. The decline in the volcanic eruption rate through time as Mars' interior cooled off would have been accompanied by a decline in the steady-state amount of CO_2 in the atmosphere. Eventually, as the volcanic eruption rate dropped to very low values in recent history, the amount of CO_2 in the atmosphere would have declined to the present-day 6 mbars (about 0.6% of the total atmospheric pressure on Earth). Impacts of planetesimals during the tail end of planetary formation would also have provided enough heat to recycle at least some of the CO_2 back into the atmosphere; the declining impact rate around 4.0–3.5 b.y.a. would have led to a decrease in the recycling rate and in the atmospheric pressure. It is not clear, though, how either the carbonate formation process or the recycling processes would have acted in the real martian environment.

It is apparent that only the largest of our terrestrial planets is undoubtedly capable of recycling CO_2 efficiently back into the atmosphere. If Mars could be placed where the Earth is today, it still would not be able to support liquid water at the surface for very long – the CO_2 would be removed from the atmosphere quickly, even the minimal greenhouse warming that it supplies would disappear, and the temperature would drop to below freezing. Warmer temperatures could not be sustained unless there were a mechanism to resupply CO_2 back into the atmosphere. Thus, planet size and geologic history play an important role in determining whether a planet will be habitable.

Another factor that plays a role in planetary habitability is the ability of a planet to retain its atmosphere. A Moon-sized object, for example, is not massive enough to hold on to any atmospheric gases for very long; they quickly would be lost to space. The Earth and Venus, on the other hand, are massive enough to retain essentially all of their gases except hydrogen and helium throughout geologic time.

What about Mars? It is massive enough to keep the major gases from escaping to space just from the velocity of the molecules through their random motion. However, various other processes can provide enough energy to eject some of the gas into space. Chemical reactions in the upper atmosphere, driven by ultraviolet light from the Sun, can give some atoms enough energy to cause them to escape from the planet completely. In particular, substantial amounts of nitrogen and oxygen may have been lost over time by this method. In addition, the solar wind may have caused a large fraction of the atmospheric gases to be lost through time: oxygen ions in the upper atmosphere have a charge that causes them to feel a force exerted on them by the magnetic field of the incoming solar wind; this force causes them to spiral around the magnetic field lines. Because the solar wind is traveling at several hundred kilometers per second, the ions will be accelerated to very high velocities. Some will be removed from Mars directly, and others will be slammed into the atmosphere at high velocity and will knock off other atoms that are present at the top of the atmosphere. This 'sputtering' process can remove a substantial amount of atmospheric gas over geologic time, possibly up to a bar of CO_2, for example.

This sputtering process works effectively on Mars because the incoming solar wind can approach the planet very closely. On Earth, the presence of a magnetic field keeps the solar wind from approaching very close to the planet, so that it does not interact with the upper atmosphere and the sputtering loss process cannot operate. Like Mars, Venus does not have a magnetic field to hold off the solar wind; the atmosphere is so thick, however, that sputtering can remove only a small fraction of the total atmosphere over time. The atmosphere is thin enough on Mars that removal of the same number of molecules represents a large fraction of the total gas available.

Mars also must have lost some gas through the force of asteroid impacts onto the surface, especially several billion years ago when the impact flux was large. The rock ejected from the surface as the impacting object forms a crater can push on the atmospheric gas and eject it from the planet. The fact that we see abundant impact craters preserved on surfaces older than about 3.5 b.y. means that this process must have acted to some extent; it may have been able to remove a substantial fraction of the gas from the early martian atmosphere. This 'impact erosion' process would have been much less effective for Earth and Venus because their greater mass makes it more difficult to remove the gas to space.

Clearly, then, the loss of gases to space will depend on planet size and on such intrinsic properties of a planet as whether it has a magnetic field. The nature of the planet itself will have a substantial influence on the availability of liquid water at its surface, and therefore, on its own habitability.

Boundaries of the habitable zone

Let's turn now to a discussion of the location of the boundaries to the habitable zone. The major influence on a planet's surface temperature at a given instant are the planet's distance from its star and its atmospheric composition. The amount of energy emitted from a star is easy to calculate or to measure; this allows for a very straightforward estimate of what the surface temperature would be in the absence of an atmosphere. It is the control of the temperature by the atmosphere, however, that determines the overall size and location of the habitable zone. We need to consider the specific atmospheric factors that control the temperature in order to determine how close a planet can be to a star before the planet becomes too warm for liquid water to exist or how far away the planet can be before it becomes too cold.

The outer boundary of the habitable zone will be determined by the farthest that a planet can be from the star and still allow water to remain liquid. Because of the ability of CO_2 in the atmosphere to raise the temperature by the greenhouse effect, this outer boundary will be farther from the star if CO_2 is present than if it is not.

Simple models of the energy balance of the atmosphere allow estimates to be made of the location of this outer boundary. For our present solar system, they suggest that a planet could still sustain water at a distance of about 1.7 A.U. from the Sun. There, a thick CO_2 atmosphere would trap enough heat for the surface temperature to rise to the melting point of ice. This distance is well outside the orbit of Mars, but allows for the possibility that more CO_2 could be present in an atmosphere than is currently in the martian atmosphere. A more conservative estimate would be based on the fact that a thick CO_2 atmosphere could begin to condense in the middle part of the atmosphere at a closer distance to the Sun. This would occur if there was nothing, such as atmospheric dust or another greenhouse gas, that could raise the temperature of the middle atmosphere enough to keep the CO_2 from condensing. This condensation could limit the amount of CO_2 that can exist in the atmosphere and, therefore, the amount of greenhouse warming that can occur. For today's Sun in our solar system, this would occur at a distance of about 1.4 A.U. (just inside of Mars' orbit); planets farther out than this might not be able to hold enough CO_2 in their atmosphere to warm the planet to the melting point of ice.

The inner boundary of the habitable zone can be estimated using the closest distance that an Earth-like planet could be to the Sun without suffering a runaway greenhouse. In our solar system today, this distance is about 0.84 A.U., about halfway between the orbits of Venus and the Earth. Venus is located well inside this distance, consistent with the presence of a thick greenhouse atmosphere (and, of course, Earth is located farther out than this distance).

A more conservative estimate can be made of the location of the inner boundary, based on a possible trigger toward a thick atmosphere that could result from the presence of only moderate greenhouse warming. On

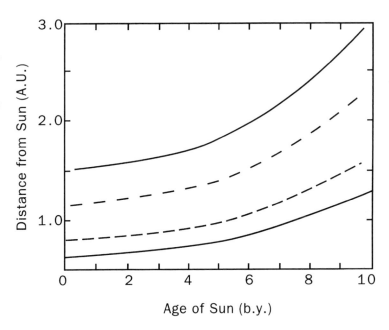

[Figure 16.2]
Inner and outer boundaries of the habitable zone (HZ) for the Sun, as a function of time since the Sun's formation. The solid lines are based on the known properties of Mars and Venus, and the dashed lines are based on more conservative theoretical arguments. The HZ moves outward from the Sun with time because the Sun's radiation output is increasing as it ages. (After Kasting *et al.*, 1993.)

the Earth today, temperatures are too cold in the stratosphere to allow much water vapor to be present there. If just a moderate additional greenhouse warming existed, the altitude of the top of the troposphere would rise and the water mixing ratio there would increase; this increase would allow water to move more readily from the troposphere into the stratosphere. Once this happens, ultraviolet light at the top of the atmosphere would break up the water and allow the hydrogen to escape to space. If this happened, the water at the surface of the planet would be lost to space relatively quickly and the CO_2 would become trapped in the atmosphere. The limit for this 'moist' greenhouse in today's solar system is at about 0.95 A.U., just inside the orbit of the Earth.

We see that the present-day habitable zone of our solar system could extend from somewhere between 0.84 and 0.95 A.U. at the inner edge to between 1.4 and 1.7 A.U. at the outer edge. Thus, there is a 'window' that is somewhere between about 0.4 and 0.8 A.U. wide in which the Earth could reside today and still allow liquid water and, presumably, life to exist.

Because the Sun's energy output has varied through time, the habitable zone has not always been in the same location. At the time of formation of the planets, when the Sun's output was 30% less than it is today, the HZ would have varied from 0.6 or 0.7 A.U. at the inner edge (using these same models of the climate) to 1.2 or 1.4 A.U. at the outer edge (Figure 16.2). Significantly, the Earth lies within the region that allowed liquid water both early in its history and continuing up to the present; this is consistent with the Earth's geological history, which provides evidence for the continuous existence of liquid water from the Earth's early history to the present (Chapter 4). If we consider the region that allowed liquid water to exist continuously for 4.5 billion years to be the relevant one, then we can determine the continuously habitable zone around our Sun. This CHZ is, of course, much narrower than the HZ is at any given instant. It might extend only from 0.95 to 1.2 A.U. (Figure 16.2).

However, this CHZ does not take account of the possibility that liquid water was present on the surface of Mars 3.5 to 4.0 b.y.a. If water was present, it appears that our models are incomplete and that other, as yet poorly understood, processes would allow liquid water to exist. If we use Mars as a constraint on the location of the outer edge of the HZ, then the HZ extended out to at least 1.5 A.U. some 4.0 b.y.a. Allowing for the increase in solar energy, the current HZ would extend to almost 1.8 A.U. (Figure 16.2).

We cannot do a similar calculation for the inner boundary of the HZ, because of the lack of knowledge of the early geological history of Venus. The strongest statement that can be made is that, as far back as we can see into Venus' past (about 1 b.y.), there is no evidence of liquid water. If 1 b.y. ago just happened to be the time at which Venus underwent a catastrophic transition to a runaway greenhouse, then we can infer that the inner edge of the HZ was no closer to the Sun a billion years ago than Venus's orbital location (0.72 A.U.); this would suggest that it was no closer than about 0.65 A.U. when the planets formed and about 0.75 A.U. today. (If Venus underwent the transition to a thick greenhouse atmosphere earlier than a billion years ago, the inner edge of the HZ would be farther out than these distances.)

Thus, it is possible that the CHZ around our Sun extends from 0.75 A.U. out to 1.5 A.U. Whether one takes the more conservative or the less conservative boundaries to the Sun's CHZ, it is clear that the zone is not extremely narrow and that the Earth is located well within it. The occurrence of life as we know it on the Earth does not appear to be an incredibly unlikely event.

Interestingly, the Sun is continuing to heat up at the present, and the habitable zone is moving outward. Possibly before another billion years pass, and almost certainly by the time another 3.5 billion years pass, the inner edge of the HZ will move outward past the Earth, and the Earth's atmosphere will change catastrophically to a more Venus-like atmosphere. The Earth has lived through more than half and perhaps as much as 80% of its habitable lifetime, regardless of how humans or other life continues to affect the planet!

Consider now the location of the habitable zone around other stars. We

[Figure 16.3]
Inner and outer boundaries of the HZ for stars of different masses (indicated as the mass relative to that of our Sun). For the more massive stars, the history is truncated when the star moves off of the main sequence of the Hertzprung–Russell diagram. (After Kasting *et al.*, 1993.)

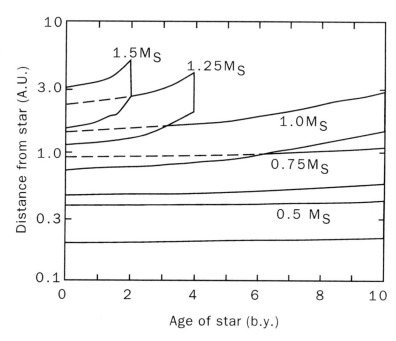

can use the same criteria regarding the response of a planetary atmosphere in order to define the location of the boundaries of the habitable zone and we can frame it in terms of the amount of stellar energy received by the planet. The amount of energy given off by a star is related to the star's mass (see Chapter 15). The life span of the star also is related to its mass, since more-massive stars are short-lived. For stars more massive than 10 solar masses (M_S), the lifetime would not be long enough even to allow planetary accretion to remove all of the debris from a stellar disk; therefore, planets around these stars would not have a climate that was stable against impacts, and an origin of life could not occur. Stars that are more massive than about 1.5 M_S have lifetimes that are shorter than about 2 b.y.; although life could originate on planets around these stars, based on analogy with the Earth it is unlikely that it would be able to evolve in 2 b.y. to the point of being detectable from space. We will not consider stars more massive than 1.5 M_S as allowing for the existence of life.

Obviously, the more energy that is emitted by a star, the farther away from the star the habitable zone will be. And, the less energy that is emitted by a star, the closer in to the star the habitable zone will be. The HZ of a 1.5-M_S star will be centered on a distance about 2.5 A.U. from the star, while the HZ of a 0.5-M_S star will be centered about 0.3 A.U. from the star (Figure 16.3). More interestingly, perhaps, the smaller stars evolve more slowly, so that the location of the HZ does not change much with time over many billions of years.

What is of the most interest, however, is whether the location of the habitable zone coincides with the location where planets are expected to form. As discussed in Chapter 14, the formation of terrestrial-like planets is thought to occur by a series of processes involving the accumulation into larger objects of dust and debris that is present in an initial disk; the gravitational accretion of the largest objects results in a small number of final planets. The final locations of the planets produced in computer simulations were located between about 0.3 A.U. and several A.U., although the presence of a gas-giant planet analogous to Jupiter could limit the number of planets that are able to exist farther out than about 2 A.U.

Clearly, the HZ around small stars will be located at the inner edge of where planets are expected to be located. For larger stars, the HZ will be at the outer edge of where planets might exist. And, for solar-mass stars, as in our solar system, the HZ coincides with the expected location of planets. Throughout this whole range of stellar masses, however, we expect terrestrial planets to form within the habitable zone. Some of these planets will be large enough to have sufficient geological activity to keep the atmosphere supplied with CO_2 and allow liquid water to be stable at their surfaces.

We can get an idea of the possible distribution of planets with habitable zones by looking at the number of stars with masses between about 0.5 and 1.5 M_S (Figure 16.4). Although stars as massive as 20 M_S or more exist, they make up a very small fraction of the total number of stars. Most stars are less than a few solar masses, and perhaps a third are between about 0.5 and 1.5 M_S. The number of smaller stars is highly uncertain, though, because they are difficult to detect and observe. In fact, our knowledge of the abundance of small stars is based only on their occurrence in the neighborhood of our galaxy that is close to the Sun, and there is little evidence that the local neighborhood is representative of the galaxy as a whole in this regard. However, we do get the sense that our Sun is typical, and that smaller stars that would have habitable regions surrounding them must be relatively abundant in the galaxy.

A major complication, however, mentioned in Chapter 15, was whether the orbital evolution of giant planets might alter the stability of the terrestrial planets. If a giant planet spirals in toward its central star, or if mutual interactions between giant planets send one deep into the inner solar system, then it is likely that the giant planet's presence will disrupt the orbits of the rocky planets. There might be solar systems around other stars

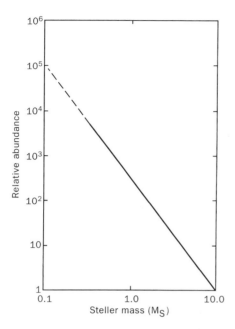

[Figure 16.4]
Relative abundances of nearby stars in our galaxy. The abundance follows a power law with an exponent of 2.5, which means that there are $10^{2.5}$ (=316) times as many stars with a mass of 1 solar mass as there are stars with a mass of 10 solar masses. The distribution is shown as dashed for small stellar masses due to the paucity of observed stars with those masses.

that have giant planets but no terrestrial planets in the inner solar system. Again, however, there is insufficient information to know whether this type of behavior might be typical.

There is another complication to the possible distribution of habitable planets. Planets that are located sufficiently close to their central stars can have their rotation slowed down by tidal forcing from the star; if slowed down enough, they could become tidally locked, always pointing the same face toward the star. This would be expected to occur for planets within the HZ for stars less massive than about 1/3 of the Sun's mass. If a planet was tidally locked, one side might become too warm, and the other side too cold, to sustain life. Of course, heat transport through the atmosphere might redistribute the heat, much as it takes heat into the polar regions on Earth during the winter.

Habitability of planetary satellites

Of course, one final complication to the location of planetary habitable zones is the question of life that might exist on planetary satellites. As we saw for Europa, a satellite of Jupiter (Chapter 13), it is possible for liquid water to occur on an object well outside of the nominal habitable zone of a star. Europa has an additional source of heat, provided by orbital interactions between the satellites and the resulting tidal forces on Europa from Jupiter. This additional heating may be sufficient to allow the ice near the surface of the moon to melt, allowing for the possible presence of sub-surface oceans that could harbor life. Clearly, the concept of a habitable zone is a simplification of the real range of possible behaviors of a planet.

Satellites of planets in other solar systems conceivably could behave in a manner similar to Europa. If multiple satellites are present, as with the Jupiter system, it is possible that they could be locked together in an orbital resonance that is capable of driving the tidal interactions. The satellites closer in to the planet will be pushed outward by tides more quickly than those satellites farther out; this means that one satellite will have the chance to catch up with the next and become locked in a resonance.

Further, there is the matter of the heat given off by the giant planet itself to consider. As with a brown dwarf object, Jupiter is giving off more energy than it receives from the Sun. The excess energy comes from heat released by the continuing cooling and contraction of the planet. For Jupiter today, 4.5 b.y. after its formation, the amount of heat left over is small and provides only a small amount of additional heating of its satellites. A brown dwarf just after formation, however, perhaps ten times as massive as Jupiter and giving off larger amounts of energy (Chapter 15), might have its own habitable zone. Certainly, it is possible that some of the substellar companions discussed in the previous chapter might have satellites orbiting them, and these satellites might have or have had liquid water at their surfaces.

Also, some of the planets or brown dwarfs that orbit close in to their central star, as discussed in Chapter 15, could have satellites orbiting around them which would meet our criteria for habitability. Even though they would have formed around a gas giant planet, they could be warmed today by the light from their central star. The types of objects out there that might be habitable almost certainly are more varied than what we can imagine today.

Concluding comments

It is not far-fetched that planets around other stars might be located at the right distance from their central star to allow liquid water to exist on their surfaces. And, if they do have liquid water, it is possible that they might have life. The chain of logic is not a unique one, however. While the analysis suggests that planets could be habitable within certain regions, the properties of the planet itself have a strong influence on habitability. While theories suggest that rocky, Earth-like planets should exist within habitable zones, these theories are relatively poorly constrained by observations; it also is possible that few planets exist within extrasolar habitable zones. Even if planets do exist there, there is no guarantee that they actually would have atmospheres, water present at the surface either as a liquid or as ice, or usable sources of energy for driving life. And, although life seems to have originated very quickly on the Earth, and therefore is seen as a likely result of simple evolutionary processes, it also is possible that life may not have originated on these planets and may not be present today. However, as the sportswriter Damon Runyon said, 'The race may not always go to the swift, nor victory to the strong, but that's the way you bet.' Clearly, our present thinking on the subject is that all of these events are possible and that they may be likely. Indeed, life may be widespread throughout the galaxy.

276 In thinking about the possible occurrence of life on planets around other stars, the most likely form of life to exist will be bacteria-like organisms. On Earth, these were the first organisms to form, they continue to exist up to the present, they are closer than anything else to the simplest organisms that can exist, and they are pervasive. More complex organisms took billions of years to evolve on the Earth, and may never evolve on other planets. Despite this, the sheer number of possible planets in the galaxy suggests to some that there might be intelligent life on some of them, and one of the most widely debated questions is whether intelligent beings exist in other solar systems. Down through the centuries, it generally has been accepted that stars would have planets, that these planets would have life, and that intelligent life would be widespread. Even the ancient Greeks thought about these concepts and accepted this idea. During the last several decades, there even have been attempts to communicate with extraterrestrial intelligent beings – by listening with radio telescopes to see if anybody is broadcasting to us, by transmitting radio waves toward other stars, and even by sending artifacts (spacecraft) with messages outside of our solar system. Although only the first of these had any real chance of success, there has not yet been a detection of intelligent life elsewhere. In fact, there is no evidence to suggest that intelligent life exists; at the same time, of course, there is no evidence to preclude it either.

Intelligent life on another planet is one of the broadest topics that we could address, so it has been saved for last. Not only does it require understanding the origin and evolution of planets and of life, but also an understanding that we do not yet possess of the nature of intelligence and consciousness itself. We can talk about the processes that might have helped lead to intelligent life on the Earth or about the physiological aspects of intelligence. However, those processes that led to human intelligence are not well understood. Thus, we can make only guesses as to how they might have operated elsewhere in the galaxy. In the end, both extremes will be left as possible outcomes – that intelligent life could be widespread, or that it could be a fluke of nature and probably does not exist elsewhere.

In this chapter, we discuss the nature of intelligence in humans and some aspects of its origin, the likelihood of intelligent life existing on other planets, and the approaches that have been taken to listening to the heavens in a search for extraterrestrial intelligence.

The origin of intelligence?

What is intelligence? As Supreme Court Justice Potter Stewart said of a very different issue, 'I know it when I see it'. The fact that we are able to ponder these questions appears to set humans apart from all other life forms on Earth. Humans exhibit self-awareness, the ability to think in abstract as well as concrete terms, and the ability to have dreams and hopes for the future. They also have the ability to affect their environment by choice; although organisms have a long history of affecting their environment (as with the production of an oxygen atmosphere some 2 billion years ago), only humans can choose to change their environment or not to change it.

Intelligence probably cannot be defined in a unique and clear way. The best definition comes about through application. If we are interested in finding intelligent life elsewhere in the universe, for example, then we must observe it directly or else establish communications with it. Because any such communications almost certainly will take place by radio (see below), intelligent life becomes defined as organisms able to build radios capable of communicating over interstellar distances. By this definition, humans became intelligent only within the last 50 years or so. Clearly, this is a faulty definition, as we have reason to believe that there has been no change in the intelligence of *Homo sapiens* since their appearance some several tens of thousands of years ago.

Although not satisfying, we also can try to quantify intelligence by examining brain size. Humans do not have the largest brain of any creatures on Earth. If one compares the brain size to the total body mass of a wide variety of terrestrial species, a clear correlation exists – the larger animals generally have larger brains (Figure 17.1). But, they do not have greater intelligence. Larger animals presumably need more of their brain power simply to control the basic functioning of their bodies. However, humans do stand out on the basic trend as having the largest brain compared to its size (that is, humans are located the farthest above the general trend in Figure 17.1). And, those animals that we recognize as being intrinsically smarter than others also stand out in this same way. For example, dolphins (and other cetaceans) stand above the general trend (Figure 17.1).

Although primates generally are intelligent animals, humans stand out (with larger brains) when compared with this group as well. It was only with the evolution of *Homo sapiens* and their immediate precursor species in the last few hundred thousand years that the brain became as large as it is. We can follow the history of the size of primate brains, along with the history of the species themselves, and learn something about the ways in which human intelligence evolved.

Modern humans evolved from earlier primates, and are thought to have originated in the grasslands of central Africa. The location of their origin is based on the dating of fossils of early humans throughout the world, with ages spread over the last two hundred thousand years, as well

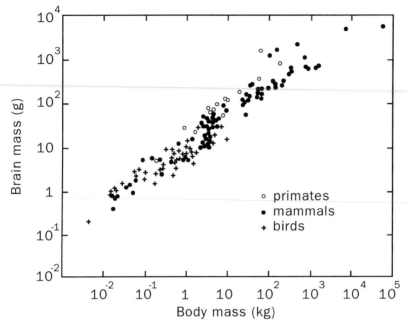

[Figure 17.1]
Comparison of brain mass versus body mass for a variety of primates, mammals, and birds. The open circle farthest above the general trend is modern humans. (After Sagan, 1977.)

as on biochemical arguments involving the evolution of mitochondrial DNA. From Africa, humans are thought to have spread throughout the world over a period of many tens of thousands of years.

The oldest hominids, the genetic line that gave rise to humans, date back about 4 million years (Figure 17.2). The oldest in this line, in the genus *Australopithecus*, stood between about 1 and 1.7 m tall and weighed between 25 and 60 kg. Even that long ago, simple tools were in use. A number of distinct species of hominids have been found that span the time from 4 m.y.a. up to the present, even though the exact line of descent is not well understood. The most recent species include *Homo neanderthalensis* (Neanderthals) and *Homo sapiens* (modern humans). Neanderthals existed as long ago as about 150 000 years and may have survived long enough to have overlapped in time with modern *Homo sapiens*. Modern humans arrived only within the last 15 000–30 000 years; we have written records of the history of civilization only for the last 5000 years or so.

What caused the development of increased intelligence in *Homo sapiens* compared with its ancestors? The brain size of hominids increased from some 400 cubic centimeters about 4 m.y.a. (in *Australopithecus afarensis*) to 700 cm³ in *Homo habilis* about 2 m.y.a. to 1350 cm³ in modern humans. During this same time period, the climate in Africa was continuing its change from tropical rain forests to a mixture of forest and savanna (containing a mixture of grass and low shrubs) and then to grasslands.

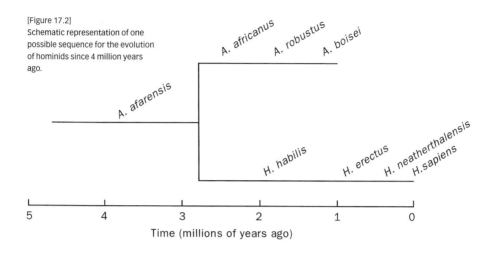

[Figure 17.2]
Schematic representation of one possible sequence for the evolution of hominids since 4 million years ago.

Time (millions of years ago)

Human ancestors went from tree-dwelling to living on the open plains. This change in habitat appears to have driven major changes in physiology, with primates developing the ability to walk upright. In turn, this freed up the hands for other functions, including increased development of tools and an increased manual dexterity.

Clearly, a major advantage in the development of intelligence was the increase in brain size accompanied by manual dexterity. For comparison, dolphins have a substantial brain, but they lack the manual dexterity to take advantage of this. And, Neanderthals actually had larger brains than humans, but still were superseded by them. Of course, intelligence is not necessary for the long-term survival of a species, as evidenced by the tremendous evolutionary success of bacteria, plants, and so on.

It is not at all obvious that there is an imperative toward development of intelligence. To explore this, we need to discuss the nature of evolution. As originally described by Charles Darwin in 1859 in his book *The Origin of Species*, species come into being by descent from earlier species. The process is what he called 'descent with modification', and involves two concepts. First, all species undergo changes or mutations. Today, we know that these can be random (copying errors in DNA, for example) or driven by external forces (such as mutations triggered by absorption of ultraviolet light by the skin), or they can express the natural variation of a given property within a species (such as height). These changes can be passed on to offspring. Second, if a change is favorable, its use will offer an advantage in reproduction to those members of the group that have it. In this way, individuals who have an advantageous change will produce more offspring than those who do not, and eventually the change will spread through the entire species.

New species evolve by the incorporation of a large number of changes. We can see this incorporation in the fossil record, by looking at the evolution of closely related species through time. Most close to home, we can see some of the changes in the line of species that led to humans – increases in brain size, evolution of the opposable thumb, the shape and orientation of the pelvis, and so on. Evolution has operated during the entire history of life on Earth. It resulted in the earliest life on Earth giving rise to a wide variety of species, from simple archaea and bacteria to more-complicated multi-celled creatures and to the wide variety of species that occupy the surface of the Earth today. Details that are being debated today, such as the role and importance of punctuated rather than gradual evolution, are not at odds with the basic nature of the overall process.

The evolution of life often is depicted in popular writings as a history of increasing complexity, with humans being the pinnacle of development. By this model, there is a natural imperative toward more-complex life and toward intelligence. Increased complexity in a species allows for increased flexibility in lifestyle, multiple choices of food supply, and so on; along with intelligence, it offers a clear advantage to a species, and natural selection should favor it. If correct, this would suggest that any planet in the universe that has life would, given sufficient time, develop intelligent life.

This viewpoint is not correct, however. First, evolution is not governed solely by natural selection operating in an otherwise unchanging environment. For example, one major complication in the history of terrestrial life has been the occasional major impact by an asteroid (see Chapter 2). The largest impacts that can occur on the Earth even during the present epoch are capable of creating a sudden and sustained change in the environment that can destroy a large fraction of existing life on Earth. And, most importantly, these changes occur within the lifetime of a single organism, so rapidly that a species cannot evolve in response to them. Rather, those species that just happen to have characteristics that would allow them to survive in the changed environment will do so. It may not be apparent ahead of time what characteristics will allow a given species to survive, and those characteristics may have evolved originally for completely unrelated reasons. The evolution of humans follows almost 600 m.y. of evolution of macroscopic life on Earth. During that time, there have been about a half-dozen worldwide, major extinction events. Each event killed a large fraction of the species on Earth, and opened up a tremendous number of new ecological niches into which those species that survived could evolve. Certainly, we, and everything else alive on the Earth today, are a product of those events.

Impacts are a random phenomenon on the Earth, based on statistical interactions between the Earth and a large number of smaller objects orbiting the Sun. If the planet were to live its history over again, the history of impacts might be similar, but it would not be exactly the same. There is no reason to expect, then, that under the circumstances of this altered history evolution would follow the same course. The nature of the species existing after 4.5 billion years would, presumably, be completely different from what we see on the Earth today.

Second, life on Earth is not defined solely by the largest creatures that we see inhabiting the lands and the oceans of the planet, even though those are the ones with which we are most familiar. Rather, it appears that the most abundant life on Earth may also be the simplest life – bacteria. In this sense, we cannot view the evolution of life on Earth as being a continuing evolution toward more-complex organisms. All the life that exists today is the equal product of 4 b.y. of evolution, whether it be the simplest bacteria, the smallest plants, humans, or the largest organisms on the planet. That is, evolution has allowed organisms to evolve to fill essentially all of the available ecological niches. The largest species – those that we see on the surface of the planet – are in some sense no more evolved than the simplest bacteria.

In this sense, we need to consider the overall diversity of life on Earth. Whether we consider number of different species or total biomass, the most abundant life on Earth is also the simplest. As we learn more about the different environments in which bacteria and archaea can thrive, and discover more places where they exist today, we rapidly are discovering that they may be the most abundant life on Earth. In this case, evolution has given us a wide variety of species of different complexity, but the most abundant ones also have changed the least from the earliest times. The macroscopic life forms represent simply a high-complexity tail on the distribution of life. Viewed in this sense, there is no imperative for the evolution of life toward more-complex forms. Instead, the imperative is for the continued existence of the simplest life forms, with various processes also allowing the evolution of some more-complex organisms as a byproduct.

Although still sparse, some evidence from the geological record supports this viewpoint. In cases where evolution of a species can be tracked in sufficient detail, there appears to be an equal chance of evolving toward less-complex forms as toward more-complex forms. That is, the evolutionary pressures that act on species do not automatically select the more-complex organism. Rather, they select the organism that will fit better into a given environment, and this can be either a more- or less-complex organism depending on circumstances.

Third, it is interesting to note that out of all of the species that have ever existed on the Earth, only one has had complex intelligence. One could argue that this is naturally the case, since only one species could be the first to develop intelligence and the existence of that species might preclude another intelligent species from ever developing. This is a false argument, however. A better comparison comes about from noting that, through the last few hundred million years, there have been a large number of species that might have evolved into intelligent creatures but did not. If intelligence is such a beneficial trait, allowing a species to compete more effectively in its environment, then one might expect that many different species would develop intelligence. Birds have existed longer than mammals, and they did not develop intelligence. Dinosaurs reigned for over 100 m.y. and there is no evidence to suggest that they might have developed intelligence. Out of all of the species that might have developed intelligence, only one did. This suggests that there is no imperative toward higher intelligence among species and that the development of intelligence is a low-probability event.

In fact, looking at the distribution of brain mass versus body mass in animals (Figure 17.1), it is clear that the imperative is toward a uniform level of intelligence, with the brain size relatively closely matched to the body size. Species with more massive brains have larger bodies that can take advantage of it, rather than smaller bodies that would allow more of the brain to be used for abstract thinking.

Viewed in this light, it would seem that intelligence is a random result of evolutionary processes. The specifics of the evolution of the high-complexity tail on the distribution of organisms probably are random. There is no guarantee that the processes would act in a way to produce an intelligent organism, and there is no guarantee that intelligence and dexterity would evolve in the same organism. From this perspective, there is no reason to think that intelligent life would evolve on a planet in another solar system. What are the odds of evolving intelligent life? Really, though, we have no way to know.

Is there intelligent life out there? The Drake equation

Given the uncertainties connected to the question of whether intelligence is or is not likely to evolve, it might seem somewhat silly to try to determine from a logical argument whether there is intelligent life in the universe. However, it is an interesting exercise that will give us insight into the para-

meters that might be important. We can estimate the number of civilizations in the galaxy with which we might communicate using a formalism
developed initially by the radio astronomer Frank Drake in the early 1960s.
The so-called Drake equation provides an estimate of the number of
civilizations by starting with the rate of star formation. Of those newly
formed stars, what fraction will have planets? Of those stars with planets,
how many planets in each system might be habitable, suitable for life? Of
those, what fraction actually have had an origin of life? Of those, what fraction might have evolved intelligent life? Of those, what fraction will want to
communicate with us? And, finally, how long will a civilization that is able
to communicate survive?

This can be written mathematically as:

$$N_c = R_s \times f_p \times n \times f_l \times f_i \times f_c \times L \tag{1}$$

where N_c is the number of civilizations in our galaxy with whom we might
communicate, R_s is the rate of formation of stars in the galaxy, f_p is the fraction of stars that have planetary systems, n is the average number of habitable planets within a star's solar system, f_l is the fraction of habitable
planets on which life arises, f_i is the fraction of planets with life on which
intelligent life arises, f_c is the fraction of intelligent species that are interested in communicating with other civilizations, and L is the average lifetime of a civilization.

Clearly, some aspects of this equation are intuitive. If a civilization only
lives for a few decades before destroying itself by war or by irreversible
damage to its planetary environment (as may still happen on Earth), then
there are likely to be very few civilizations in the galaxy at any one time. If
civilizations live for billions of years, many more are likely to be exist.

It is instructive to go through each factor in this equation and choose
some representative numbers in order to see what the results are. R_s is relatively straightforward, but each of the other parameters has some uncertainty and some of them have considerable uncertainty. So, we will take
both an optimistic and a pessimistic viewpoint, and then look at the range
of possible answers.

R_s: There are approximately 10^{11} stars in our galaxy. Given the age of the
galaxy of about 10 b.y., this means that the average rate of star formation
has been around 10 stars/year. Why not just take the current number of
stars in the galaxy, and continue from there? If we look at the existing stars,
we are seeing stars of all ages from 0 to 10^{10} years. This gives us a biased
sample of stars. For instance, we are seeing only young massive stars rather
than old massive stars, because the old ones that formed early in the
galaxy's history have already died. Rather, we wish to look at the steady-

state occurrence of intelligence, so we look at the steady-state rate of star formation.

f_p: What fraction of stars have planets? Based on the occurrence of protoplanetary disks (Chapter 14), it could be as many as half. Since not all disks will of necessity form planetary systems, though, the number could be much smaller. Additionally, stars in binary or multiple star systems may not be able to develop planets or their planets may not be habitable, and more than half of the stars are in such systems. Let's take a pessimistic value of 0.01 and an optimistic value of 0.3.

n: On average, how many planets in each solar system will be habitable? Based on the theories of planetary formation (Chapters 14 and 15) and the habitability of planets (Chapter 16), we expect that habitable planets might be common. In our own solar system, we saw that between one and three planets have been habitable (certainly the Earth, maybe Mars or Venus). An optimistic number might be three, and a pessimistic number might be one. Of course, one could be truly pessimistic and imagine that the oddball giant planets (Chapter 15) whose orbits evolved through the inner parts of their solar systems would have destroyed any terrestrial planets; in this case, perhaps a pessimistic number might be, on average, one habitable planet in every hundred solar systems, or 0.01.

f_l: On what fraction of habitable planets does life form? If life is as much a natural consequence of simple chemical processes as we think it might be, this could be close to 100%. On the other hand, life might be much more of a fluke, in which case we have no basis to choose a pessimistic number. For the sake of discussion, let's choose a pessimistic fraction of one in a million (10^{-6}) and an optimistic fraction of 100%.

f_i: On what fraction of those planets with life does intelligence develop? Again, we have little basis for choice, and the numbers could be anywhere from one in a million (or less) to 100%. Of course, a star has to live long enough to allow intelligent life to evolve. This evolution took about 4.5 b.y. on the Earth, and a sizable fraction of stars do not live this long. Let's choose extremes of 1 in 2 (0.5) and 1 in a million (10^{-6}).

f_c: What fraction of intelligent civilizations will want to communicate? If they are all like us, this will be close to 100%. If they aren't, then we can't say. Let's arbitrarily choose 100% here.

L: What is the lifetime of a civilization? As a civilization with the ability to communicate, humans have existed (so far) for only a few tens of years. If we destroy ourselves in the near future, the lifetime could be as short as 100 years. If civilizations can figure out how to survive without destroying themselves, and if they can survive the impacts that might continue on their planet, they might be limited only by the lifetimes of their stars. Our own Sun has a lifetime of 10 billion years, of which about half has passed **285** already. Of course, the increasing solar output with time may make the Earth uninhabitable after only another one or two billion years (Chapter 16). Let's choose extreme values of 100 years and 1 billion years.

Multiplying these numbers together gives us an estimate of the number of civilizations that currently exist in the galaxy with whom we could communicate. The pessimistic view is that $N_c = (10) \times (0.01) \times (0.01) \times (10^{-6}) \times (10^{-6}) \times (1) \times (100) = 10^{-13}$. In this case, we would need to search 10^{13} galaxies before finding another intelligent civilization; it would be likely that we would be the only intelligent species in the universe. The optimistic view is that $N_c = (10) \times (0.3) \times (3) \times (1) \times (1/2) \times (1) \times (10^9) = 5 \times 10^9$. In this case, most stars have planets, most planets have life, and most planets would evolve intelligent life; a large fraction of the stars in our galaxy could have intelligent life in their systems.

Clearly, such a wide range of answers limits the utility of using the Drake equation to learn anything new about intelligent life in the universe. In other words, the Drake equation is just a mathematical way of saying, who knows? However, it does allow us to focus on those issues that have the largest uncertainties: Does life occur on every planet where it is possible? Does intelligent life evolve as an imperative? How long can a civilization last?

Why aren't they visiting us?

A couple of additional interesting approaches can be taken regarding the occurrence of extraterrestrial intelligence – Fermi's paradox and the occurrence of UFOs.

The physicist Enrico Fermi asked in the 1940s 'Where are they?', referring to extraterrestrial intelligent species. His argument followed this logic: If intelligent life exists, it rapidly (over perhaps only thousands of years) would develop the technology that would allow it to travel between the stars. Even at speeds much less than the speed of light, speeds that we can imagine are achievable, travel between the stars is feasible. It might take many generations for a spacecraft to reach another star, but it would not be impossible. It is hard for many to believe that a civilization would not want to travel between stars if it could do so. Suppose a civilization sent spacecraft out to the nearest several stars. The trip would take only hundreds or perhaps a thousand years at speeds of 1–10% of the speed of light. Once there, if a planet existed that could be colonized, it would take less than

286   1000 years (only 30–100 generations for humans) to build up a new civilization and again develop the ability to send spacecraft to the nearest stars.
Spacecraft easily could be sent toward all of the nearest stars. At relatively
short intervals, then, a civilization could expand out into the galaxy, colo-
nizing all of the planets that were found to be habitable.

At a travel speed of 1% of the speed of light, and allowing a suitable time
on each planet before going on to the next, 1000 years for example, a
civilization would overrun the galaxy in less than a few tens of millions of
years. This is such a short time compared with the age of the galaxy, or even
the age of the Earth, that Fermi wondered, where are they? Why have we
seen no evidence that our planet has been visited by such a civilization? It
would be extremely unlikely that Earth is the first civilization to appear in
the galaxy. Therefore, there must be no other intelligent life in the galaxy.
This logical argument is known as Fermi's paradox.

Of course, there are a large number of solutions to Fermi's paradox. For
example, extraterrestrials could have visited the Earth but not interfered
with its development. They could be waiting to see how we will develop and
when we will join the 'Galactic Club'. Or, the period of exploration and colonization of the galaxy, during which a civilization would have a desire to
expand into the galaxy, may be short-lived and not allow for the complete
colonization of the entire galaxy by a single civilization. Or, there may be
some fundamental reason why interstellar travel is not feasible – it might
be technically more difficult or more expensive than we currently imagine.

Or, some argue, extraterrestrials could be visiting the Earth on a regular
basis today, in the form of space aliens and UFOs ('unidentified flying
objects'). A substantial fraction of the population believes that UFOs are
real and represent visiting extraterrestrial intelligent species. The usual
scientific approach is to dismiss UFOs on the basis of a complete lack of
any physical evidence. At least one person has argued that an intelligent
extraterrestrial species that wanted to reveal itself to Earthlings would
follow exactly the pattern seen in UFOs – gradual exposure to the population, increasing acceptance of their existence, and, finally, revelation of
their true form and existence.

Searching for intelligent life

Finally, we come to the question of the active search for extraterrestrial
intelligence, which often goes by the acronym SETI (with extraterrestrial
intelligence being abbreviated ETI). Guiseppe Cocconi and Philip

Morrison argued in a seminal paper published in the journal *Nature* in
1959 that it was possible to use radio waves to search actively for signals
from intelligent species. (And, as an aside, even the earliest radio experimenters, Nikola Tesla and Guglielmo Marconi, thought that they were
detecting radio broadcasts from Mars during the first two decades of this
century.) In the light of the above discussion of the evolution of intelligence, it is interesting to note that the strongest supporters of SETI are the
astronomers and physicists. They come from the background of radio
astronomy and the technical ability to communicate between stars. The
strongest detractors are the evolutionary biologists. As a group, they
believe that intelligence must be such a rare phenomenon that SETI is not
likely to produce any positive results. However, the final sentence of the
paper by Cocconi and Morrison sums up many of the issues connected to
the search for extraterrestrial life: 'The probability of success is difficult to
estimate, but if we never search the chance of success is zero.' They were
arguing that we know so little about intelligence and evolution that the
question of extraterrestrial intelligence should be treated from an experimental rather than a theoretical viewpoint – in order to find out what is out
there, we should look.

How would one go about trying to communicate with an extraterrestrial civilization or to see if one was trying to communicate with us? One
could easily imagine a variety of ways of sending information back and
forth between planets around different stars, such as by sending spacecraft directly, by transmitting visible light (either as large-scale beacons or
as narrowly focused laser beams), or by transmitting radio waves. For a
variety of reasons, radio transmission seems to be the most effective way of
communicating. Spacecraft take a very long time to travel between stars,
and there is no guarantee that they could continue to operate for the billions of years that it might take to contact an up-and-coming civilization.
Visible light is an efficient means of communication, but is easily absorbed
by dust in the interstellar medium. In addition, detecting a visible light
pulse above the natural emission of a star would be difficult, due to the
intensity of the visible light given off by the star.

Radio waves, on the other hand, pass relatively freely through the interstellar medium, are easy to detect above the threshold of a star's natural
radio emission, and, perhaps most importantly, are a means of
communication that even a fledgling intelligence such as ours could construct and operate easily. At the same time that our society was just beginning to send spacecraft to the nearest planets in our own solar system, we
were able to turn giant radio telescopes to the heavens and detect any messages that might have been sent to us from across the galaxy.

Of course, radio communication has its own difficulties, as well. Although it is easy to point a radio telescope at either a nearby or distant star and to look for energy being beamed toward us, there are a very large number of different radio frequencies that could be used for transmissions. And, there are a large number of stars that would need to be looked at. How can one choose the best way to search?

One way is to choose a 'natural' frequency of transmission, one that would be a constant throughout the galaxy and which other civilizations also would choose. Cocconi and Morrison suggested that the most natural frequency corresponds to the wavelength at which hydrogen, the most abundant element in the galaxy, emits energy. Emission at this wavelength, 21 cm, was the first energy detected from the galaxy, and it represents a potential natural beacon. Unfortunately, hydrogen also absorbs radio waves at this wavelength, so that transmissions would not travel more than a few tens or hundreds of light years (one part in a thousand of the distance across the galaxy). Few people felt that we would be so lucky as to find extraterrestrial intelligence in our own backyard, within only a few tens of light years of our own Sun; a search of more distant stars would be necessary.

What wavelength would be most suitable? At radio wavelengths shorter than a couple of centimeters, gases emit and absorb energy very efficiently. Our own atmosphere will block out much of the energy coming from other stars, and emission from interstellar gas clouds also will complicate the picture. At wavelengths much longer than a couple of tens of centimeters, natural emission from stars increases dramatically due to the acceleration of electrons around stellar magnetic fields. The region between these extremes may be the best place for a search. If the 21-cm H line is one natural frequency for a search, then so may be the nearby 18-cm line at which the abundant interstellar gas OH emits energy. And, many people noted that H and OH together make water, which appears to be a requirement for life. They reasoned that ETI would also see these as 'magic' frequencies, and would concentrate their broadcasting or their own search near these wavelengths. In essence, they argued, the region between these frequencies represents a natural 'water hole', around which civilizations might congregate (Figure 17.3).

There still are a large number of frequencies between 18 and 21 cm. Depending on the bandwidth chosen (the range of wavelengths or frequencies searched as a single unit), up to 10^{10} different frequencies would have to be examined. Modern receivers, though, have the ability to make simultaneous measurements at over a billion distinct and separate frequencies, somewhat lessening the problems.

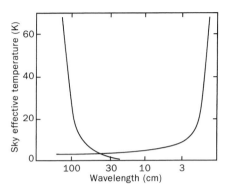

[Figure 17.3]
Schematic diagram of the background noise throughout the radio region of the spectrum. The increased noise on the left is due to radiation from the galaxy. That on the right is due to emission from the Earth's atmosphere. Some scientists have suggested that the region between 18 and 21 cm wavelength is the ideal location for a search for signals from extraterrestrial intelligent beings; these wavelengths correspond to emission by OH and by H, respectively, so that this regions is the interstellar 'water hole'.

And, there still is the problem of identifying what it is about a received signal that would be indicative of an extraterrestrial intelligence. This is exemplified by an incident in the early days of radio astronomy. In 1967, Jocelyn Bell, then a graduate student at Cambridge University in England, and her advisor Antony Hewish, set out to use an array of antennas that could measure the radio emission from quasars. Quasars are large extragalactic objects emitting tremendous amounts of energy. Their goal was to measure the 'scintillation' in the signal and to use this to try to determine the distance to the objects. Bell noticed some noise in the signal as it was recorded by a pen on a strip chart. When the noise turned up again months later, in a similar part of the sky, they recorded it at higher time resolution. Amazingly, they discovered that the noise consisted of pulses of radio energy that appeared at a surprisingly constant rate of about once per second. This was almost exactly the type of signal that might be expected from an ETI. Bell and Hewish discussed 'going public' with an announcement of the discovery of signals from intelligent species in the galaxy, but held off. They eventually catalogued several similar sources, and gave them the designations 'LGM-1', 'LGM-2', and so on, for 'little green men', still thinking of the possibility of intelligent life. Eventually, it was demonstrated that these radio emissions were a natural phenomenon, resulting from the rapid rotation of neutron stars, and these objects now are called pulsars (see Chapter 15). Clearly, for a signal to be classified as intelligent, it must be firmly demonstrated *not* to be the result of any possible natural phenomenon, including those that have not been detected previously!

A radio signal from an intelligent species probably would have a very narrow bandwidth, rather than being spread out over a wide range of wavelengths. If it were a simple beacon, designed simply to announce 'We're here!', it might be a sequence of pulses of different lengths representing whole numbers or prime numbers. Or, a message could be a sequence that could be assembled into a picture, constructed, for instance, as pulses representing a two-dimensional image (Figure 17.4). The signal probably would repeat itself on a regular basis, to demonstrate that it was a real signal. Unfortunately, a real ETI signal also might wander around in frequency (as the planet of origin moves around its star, due to the Doppler shift of the signal), it might not repeat itself if the beings were pointing their antenna at different stars on different days, or it might be based on a logic that is too foreign to us to be recognized as an intelligent signal.

The first search for extraterrestrial signals was performed in 1960 by radio astronomer Frank Drake, using an 85-foot antenna in Green Bank, West Virginia. Drake had decided to target two nearby stars, Tau Ceti and Epsilon Eridiani, both about 10 light years away, and he chose the hydrogen emission wavelength of 21 cm for the search. Their first target, Tau Ceti, showed nothing out of the ordinary. Their second target, however, blasted forth with a repeating signal of pulses at a rate of about eight per second. It was hard to imagine that they would be so lucky as to get a positive signal on their first night of observations! The scientists at the controls began looking for possible sources of interference (a prudent thing to do in an endeavor such as this), either within their instrument or from nearby sources. After about five minutes, the signal stopped and has never repeated itself. In all probability, it was the result of some unknown interference, perhaps from a nearby car or airplane engine.

Since the first observations, there have been several tens of different programs to look for extraterrestrial life. Today, there are several ongoing SETI observational programs. One of the most substantial is known as the High Resolution Microwave Survey (HRMS), renamed from 'SETI' because of the difficulty of getting Congressional funding to look for 'little green men'. This program was initiated by NASA and run by the SETI Institute in northern California, but the funding was taken over by private sources when Congress eliminated its support in 1993. In addition, a group at UC Berkeley has been operating the Search for Extraterrestrial Radio Emissions from Nearby Developed Intelligent Populations (SERENDIP), which makes use of a receiver that can operate simultaneously at over forty trillion different frequencies. This program operates in a 'piggy-back' mode, making measurements wherever other investigators decide to point the radio telescope as they study the galaxy and the universe. In addition, a group at Harvard University, funded by The Planetary Society, operates Project META (Megachannel Extraterrestrial Assay).

Several different approaches are being taken with these searches. First is an all-sky survey, designed to map the entire sky relatively quickly.

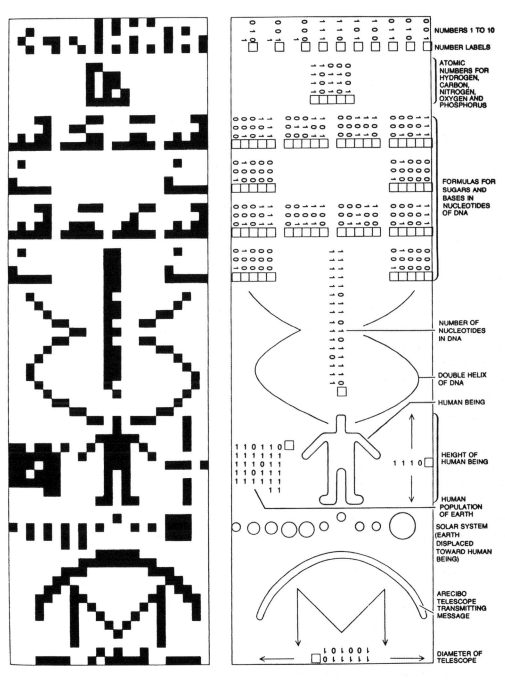

[Figure 17.4]

An example of a two-dimensional image that can be sent as a string of bits by a radio transmitter. This particular image is the one sent by the Arecibo radio antenna in Puerto Rico as a part of its inauguration in 1974. The length of the string of bits is the product of two prime numbers, so that it can be broken into a 2-D image that contains patterns that are recognizable as containing information; here, the boxes are either shaded or left blank, depending on whether the bit was a 1 or a 0. The right-hand side shows the interpretation. (From Sagan and Drake, 1975, and reproduced with the permission of *Scientific American*.)

Second is a search of nearby stars, which assumes that any signals from them would be easier to detect than from farther away. Finally, there is the search of specific candidate stars, chosen on the basis of their stellar type, age, and likelihood of having Earth-like planets. To date, no signal has appeared that has repeated itself and shown characteristics of being sent by extraterrestrial intelligence.

Concluding comments

So, where do we stand on the possibility for the existence of intelligent life elsewhere in the universe? There is no consensus as to whether human intelligence is unique or if intelligent life is likely to be widespread throughout the universe. Although many evolutionary biologists believe that intelligence is so unlikely that we are most likely alone in the galaxy, that view runs counter to another deeply held belief of many scientists – that there is no reason to think that we are special in the universe. This 'principle of mediocrity' leads scientists to believe that processes have not happened on the Earth that cannot and do not happen elsewhere in the galaxy. If intelligent life is so unlikely that we are unique, then it is hard to imagine that it would arise even once.

The bottom line is that we have large quantities of speculation and small quantities of data. The general nature of science – and of geology, planetary science, biology, and astronomy especially – is one of relying heavily on empirical tests of theories. We cannot determine the nature of planets, the evolution of life on Earth, or the history of galaxies or the universe just on the basis of physical laws alone, but we need observations to tell us what evolutionary paths have been followed. And, in the search for extraterrestrial intelligence, we also need observations to tell us what paths have been followed.

Is there extraterrestrial life? This question has been around for as long as we have had written records of our history, and probably longer. It is one for which we do not have an answer today. But, we do have a tremendous amount of relevant information. Briefly, we have every reason to believe that there could be life on other planets. In this chapter, we will summarize the logic that has led us to this conclusion. We also will discuss what the significance to society would be of finding extraterrestrial life, whether it be simple bacteria or intelligent life. And, we will summarize the current status of the search for life in the universe – what ongoing and future exploration programs will lead to us having more information that eventually could give us the answer about whether there is life elsewhere.

Is there life on other planets?

The argument suggesting that there is life on other planets is very simple and straightforward. It begins with life on Earth as that is, of course, the only life that we know about. The fossil record and the genetic record tells us that life on Earth is very old, having originated somewhere before 3.5 billion years ago and possibly before 3.85 b.y.a. The oldest known life forms were at once both very simple and remarkably complex. The simple aspect is that they consisted of microscopic single-celled bacteria and archaea. These were much less complex than the macroscopic life forms that exist today and that have existed for around a billion years. Even the oldest life forms that we can find in the fossil record, however, dating back to 3.5 b.y.a., are very complex – they are very sophisticated organisms that relied on DNA and RNA to transfer genetic information, on ATP to store energy in a usable and accessible form, and, for some of them, on photosynthesis to get access to energy. These organisms are much more complicated than we imagine the first life forms would have been.

Life did not exist, however, for one-half to one billion years following the formation of the Earth. The earliest half-billion years on the Earth were marked by the continued influx of impacting objects left over from the formation of the planets. Even today, these objects are capable of dramatically affecting the terrestrial environment when they collide with the Earth. Back then, when larger objects were present, they were capable of sterilizing the Earth completely. There were an estimated five or so impacts subsequent to about 4.5 b.y.a. that were capable of completely sterilizing the Earth's surface. The early environment on the Earth, therefore, was not conducive to the continuing existence of life, and the earliest life may not have been able to grab a foothold until sometime after around 4.2 or 4.0 b.y.a. Thus, life may have taken as little as a hundred million years, and no more than about a half-billion years, to form once the environment became sufficiently clement to allow it.

The rapidity with which life originated on the Earth has important implications for the process of forming life. It tells us that the formation of life is not a difficult event, but, rather, is a relatively straightforward consequence of natural events on the planet. This is consistent with our current view that life originated through chemical and geochemical processes, starting with organic molecules in a wet surface or near-surface environment and using energy from some chemical source to build them into more complicated molecules. Even though we do not yet understand the specific processes that led to the origin of life, we can easily see that simple chemical processes can lead to more complicated molecules and, eventually, to life.

On Earth, we imagine that life really needed only a few key ingredients to get started. Liquid water is one such substance. It is difficult for us to imagine that life could exist without liquid water. Even with this requirement, we still see that life could be widespread throughout the universe; if we allow for the possibility that a different liquid also could hold the key to life, then life could be even more widely distributed. The second ingredient is access to the necessary biogenic elements, such as C, H, O, N, and so on. This is not a very limiting factor, though, since we expect these elements to be very widespread throughout the universe, to be incorporated into planets during their formation, and to be readily available at the surfaces of geologically active planets.

A source of organic molecules was required for the origin of life on Earth, and presumably would be required for life elsewhere as well. On Earth, organics could have come from one or more of several different sources. These include the Earth's atmosphere, where they could form from energetic processes such as lightning in a slightly reducing atmosphere (the so-called Urey–Miller process); hydrothermal vents at the bottoms of the early oceans, heated by the extremely active volcanism that would have been present then, where organics could form by a chemical slide toward equilibrium as very hot water cools off once injected into the oceans; or from organic molecules that were present in dust and planetesimals accreting onto the Earth. Most likely, all of these sources contributed to the prebiotic supply of organic molecules.

Finally, a source of energy is needed to power life. The energy causes the molecules in the environment to react, moving them out of their natural state of chemical equilibrium. As they move back toward equilibrium, they

can release chemical energy to power other chemical reactions, thereby   **295**
providing usable energy for biota. Again, there are several possible sources
of energy, including sunlight (especially the energetic ultraviolet light that
could have penetrated all the way to the Earth's surface in the early periods
before there was significant ozone), lightning in the atmosphere, or geochemical energy obtained from geothermal heat in water circulating
through hydrothermal vents. All of these energy sources were available,
probably in abundance. There is no need for the energy sources that drove
the earliest life to be the same as those that power life today; thus, the complicated chemical mechanisms that drive photosynthesis did not have to
be present in the original life.

We expect that, under these conditions, the formation of life was relatively straightforward. We also expect that life could originate and continue to exist any place where similar environmental conditions are met.
This could mean elsewhere in our solar system, or on planets around other
stars. A search for life, therefore, is almost tantamount to a search for the
basic environmental conditions in which life could exist.

Elsewhere in our own solar system, we immediately think of Mars as a
possible abode for life. There is abundant geologic evidence on the
martian surface to indicate that liquid water has played an important role
in shaping the surface throughout time. The evidence suggests that water
was relatively stable at the martian surface during the first half-billion
years recorded in its geology (from about 4.0 to 3.5 b.y.a.). If correct, this
might suggest that life could have originated on Mars' surface at that time.
Subsequent to 3.5 b.y.a., however, there also is abundant geological evidence for the continued presence of water. At this later time, however, the
water was not stable as a liquid at the surface, except perhaps intermittently. Rather, water was present deep within the crust, and was
released to the surface in catastrophic floods only occasionally. Within the
crust, however, the water would have been available to support either an
origin of life or its continued existence if it had originated earlier. And,
within the crust, there was an abundant source of energy from the volcanic
activity that has persisted throughout most or all of martian history and
from chemical weathering of the minerals comprising the crust. Life could
have originated at the surface on early Mars or in the deep subsurface at
any time, and life could exist today. If life is present today, it likely would be
either deep beneath the surface where water could exist as a liquid (several
kilometers, perhaps) or exposed at the surface in any transient vents where
hot, volcanically heated water is released at the surface. Although there is
some evidence to suggest that there might be fossils from organisms
within meteorites that are from Mars, this evidence is very controversial

and is not yet generally accepted. Significantly, even if this meteoritic evidence is wrong, the basic argument regarding the possibility of life on Mars will not change; this is true even though the meteorite findings appear to have reinvigorated the interest in searching for life on Mars.

After Mars, other suggestions for an abode for life become more speculative. Life could have arisen on early Venus, when the Sun was dimmer, temperatures were lower, and the planet might not yet have undergone a transition to the present thick, hot, greenhouse atmosphere. Of course, any evidence of an early Venusian biosphere would have been long since obliterated.

Life also could exist on Europa, a satellite of Jupiter, living in a possible ocean of water that may lie buried beneath the surface covering of water ice. There, melting of the ice would result from tidal heating generated by Jupiter tides, triggered by gravitational interactions with Io as they both orbit around Jupiter. If there is an ocean, tidal heating and decay of radioactive elements would provide a substantial source of geothermal energy that might be tapped by living organisms. Although there are exciting images from the Galileo spacecraft that suggest that liquid water has been present beneath the surface of Europa, there is no certain evidence for the existence of an ocean. Life conceivably could exist on Io, as well; there, abundant energy is available through the tidal heating, although there is no evidence for water of any sort.

Life also might have existed on Titan, a satellite of Saturn. This is much more speculative, because temperatures today are much too low to allow plausible life forms to exist. There might have been liquid water early in Titan's history, however, with the heat to melt the abundant water ice being provided by large impacts during the end of the satellite's formation. Even without active biology, Titan represents an interesting exobiological laboratory, where organic chemical processes occur even today in a manner similar to what might have occurred on the early, prebiotic Earth.

As we move outside of our own solar system, the prospects for finding environments suitable for life become still more speculative. As of today, we do not know of a single planet around another star that provides an appropriate habitat for life. However, this does not mean that we have no information on the subject.

A theory of how planetary systems form as a natural byproduct of the formation of stars has been developed. This theory is based strongly on the conditions that we see in our own solar system. However, it also is based on astronomical observations of star-forming regions in the galaxy, interstellar clouds of gas and dust, actual disks of gas and dust that occur around young stars, and, now, direct detections of giant planets and brown dwarfs around other stars. As a result, there is strong reason to believe that this theory might be more general than if it were based only on our own solar system.

Planets are thought to form from the collapsing gas and dust that eventually become a star. As the cloud collapses due to the pull of its own gravity, it will begin to spin faster due to the conservation of angular momentum. Because it is spinning, not all of the material can collect into a single central ball that becomes the star, and some of the matter will stay behind as a disk around the protostar; this disk consists of dust grains and gas, in orbit around the newly formed star. The dust will begin to accumulate into larger objects, first by sticking together due to electrostatic forces, and later by gravitationally attracting other nearby objects. Eventually, these planetesimals become large enough to accumulate into a small number of individual protoplanets, each of planet-sized proportions. Only the rocky material can accumulate at the relatively high temperatures that occur close in to the star. Farther out, where temperatures are cooler both from the lesser compression of the protoplanetary disk and from the greater distance from the central star, water ice also will condense and accumulate. The greater mass available due to the presence of water ice allows more massive planetary cores to accumulate, and these then can begin to attract the gas that also resides in the disk. The gas accumulation allows giant planets, similar to our own Jupiter and Saturn, to form.

These processes are thought to be relatively general, allowing the formation of planetary systems that might look much like ours. Numerical simulations of the formation of rocky planets suggest that our inner solar system might be typical, consisting of a small number of planets in well-spaced orbits. If so, this suggests that habitable planets might be relatively common – there will be a significant likelihood of finding a planet at just the right distance to allow liquid water to exist. And, a relatively wide range of distances from the central star would allow this. In our own solar system, the habitable zone might extend from almost as close in to the Sun as Venus to almost as far away from the Sun as Mars. There is at least one habitable planet in our own solar system, and possibly as many as three or four more that might have been habitable at one time or might still be habitable today.

Significantly, we are now able to detect planets that are orbiting other stars, and we are finding that they are relatively abundant. For the most part, we cannot yet detect Earth-sized planets, only gas giants. It is hard to estimate what fraction of stars might have planets. While as many as half of the young stars have protoplanetary disks that may lead to planets, less than 10% of the more mature stars that have been examined seem to have gas giant planets. Unfortunately, the statistics for Earth-like planets are not known and cannot be determined from the available information; even for gas giants, such a small number of stars has been examined that the statistics may not be reliable yet.

The planets that we *are* detecting are providing new information on how planetary systems evolve. For example, gas-giant planets have been discovered that are much closer in to their star than was expected. These almost certainly would have to have migrated in toward the star from farther out, a process that would have devastating results for any terrestrial planets.

Given that planets do exist, however, we imagine that there must be abundant rocky planets, and that many of these will be within their star's habitable zone. This means that liquid water probably will be abundant on planets in our galaxy. If life really is able to form as easily as we think it can, under the proper conditions, then it is likely that life is rampant throughout the galaxy. Of course, life is much more likely to take the form of bacteria-like organisms than of larger, more-complex organisms. While we expect that evolution will occur on other planets, and that more complicated forms of life could exist, it seems most likely that life on other planets will be like the simplest life forms on Earth – those that have existed for the longest time and in the most varied environments, and those that may be the most abundant forms of life on Earth – bacteria.

Does this mean that intelligent life does not exist elsewhere in the galaxy? It is hard to say. On the one hand, some scientists suggest that increased intelligence offers such a tremendous benefit to an organism that it must be a highly likely outcome of evolution on any planet, given sufficient time. On the other hand, intelligent life on the Earth is the outcome of a random series of evolutionary processes, and there appears to be no natural imperative either toward more-complex organisms or toward more-intelligent organisms, so it may be that intelligence is a rare phenomenon.

Whether we are speaking of bacterial life or intelligent life, however, we have insufficient evidence today to know for certain whether extraterrestrial life exists. Although the present discussion is based on solid observations of life on Earth and the nature of the universe, the application to the question of life elsewhere so far is purely theoretical. It is only through the continued exploration of our home planet, our solar system, and our universe that we can hope to find fundamental solutions to the questions surrounding the existence of life.

Extraterrestrial life and the human condition 299

The exciting aspects about the question of the existence of extraterrestrial life, though, is that it is a question that has an answer, and we have the capability of finding that answer. What would it mean to us as a society to discover that there is extraterrestrial life? Conversely, what would it mean to search for it, not find it, and realize that life on other planets may not exist?

Human thought about our role and position in the universe has changed substantially throughout time. Prior to the 1500s, most people thought that the Earth was at the center of the universe. Although popular lore today says that when Columbus set sail for the New World in 1492 he was bucking the trend that the Earth was flat, that is not the case; most educated people knew then that the Earth was round, even as they also knew that it was at the center of the universe.

Several important events over the last few hundred years have changed how we see our place in the universe. In the mid-1500s, the Polish astronomer Nicholas Copernicus suggested that the Earth revolved around the Sun, rather than the other way around. This view seemed to simplify the nature of heavenly interactions. If the Sun and planets revolved around the Earth, they would have to be moving in complicated trajectories. They could not be moving attached to spheres around the Earth, because they appeared to change their direction of motion in the nighttime sky. Rather than moving on spheres, they would need to be attached to smaller spheres moving on larger spheres; and, as more observations were made, there was the need for still smaller spheres on these. This 'spheres upon spheres' theory proved to be too awkward. Alternatively, if the planets, including the Earth, all revolved around the Sun, then the retrograde movement seen for the planets could easily be explained by the combined movements of the Earth and the planets.

The recognition that the Earth was not at the center of the universe was one of the most fundamental discoveries in the history of human civilization. The Earth and humans became mere players in the universe, rather than its centerpiece. Because of this tremendous change in world view, Copernicus' views were not embraced by the Church; the history of his persecution is well known. This change away from an Earth-centered world does little to change the daily activities of very many people even today, however; unless one is engaged in sending spacecraft to other planets, it just does not affect our everyday activities very much. But, it has a profound effect on our view of the world around us.

A second major change in our world view came about in the mid- to late-1800s, when Charles Darwin and Russell Wallace recognized that new species of organisms on the Earth were created from existing ones by mutation and natural selection. The resulting 'Darwinian revolution' was no less profound than the earlier Copernican revolution in terms of how people fit into their world. The development of humans now could be traced from the existence of earlier species. The lineage went back not just a few million years to early primates, as discussed in the previous chapter, but could be traced back essentially throughout the entire history of life to the earliest history of the Earth. Humans no longer could be considered as special creatures on the Earth, but just as one of millions of species that resulted from four billion years of evolution following the origin of life.

This discovery is one that a few religions have had a hard time accepting, and the controversy continues even today. In 1996, the Pope, as head of the Roman Catholic Church, finally made a statement acknowledging that the evidence for the occurrence of evolution was so overwhelming that it no longer could be considered as only one theory among many. There are still religious groups that do not accept this view, however, including the fundamentalist Christians and the creationists. They argue that the scientific evidence for evolution is weak and results from a complete misreading of the geologic record. Their preferred history is based on a literal reading of the Bible, with the Earth, the entire geologic record, and all life (both living and extinct) existing only within the last 10 000 years. Some creationists are moving in the direction of accepting evolution, but they insist on a 'directed' evolution, in which the hand of God guides the changes in species so as to result in humans as the ultimate end product of this process.

A third change in our world view was much less noticed and perhaps less significant than the first two. In the 1920s, astronomers recognized that the Sun was situated within a spiral galaxy of stars, and, in fact, was located toward the outer edge of one of the galactic arms. Not only was the Sun not at the center of the universe, but it was located in the outskirts of an average galaxy. This discovery did not have the same impact as the earlier discoveries perhaps because it simply reinforced the view that our planet and our star were not central players in the universe.

The recent discovery of planets in orbit around other stars may stand as an event of significance equal to these earlier events. Although the existence of planets was suspected based on observations and theories of how stars and planets form, the significance of their direct detection cannot be overstated. They tell us that our solar system is not unique in the universe, that planets are common and that there may be innumerable Earth-like planets in our galaxy.

Where would the discovery of extraterrestrial life fit in? Actually, it may be a case similar to that of the discovery of planets. The fundamental revolution in thought may already have occurred, perhaps spread out over the last 30 years. There has been a gradual recognition that life probably formed on Earth through chemical processes acting within the ambient environment, that a similar environment on another planet could result in the formation of life there, that there are several planets and satellites within our own solar system that have the potential to harbor life, and that planets with appropriate environments may be widespread throughout the universe. Of course, expecting that life will be found on other planets is different from actually finding it.

Finding non-terrestrial life would be the final act in the change in our view of how life on Earth fits into the larger perspective of the universe. We would have to realize that life on Earth was not a special occurrence, that the universe and all of the events within it were natural consequences of physical and chemical laws, and that humans are the result of a long series of random events.

Such a discovery would not necessarily have major ramifications for most religions. They easily can tolerate the occurrence of life elsewhere in the universe. This is as it should be, because religions operate on faith, based not on what actions or phenomena occur in the physical universe but, rather, on what exists away from the physical reality. As one astronomer put it, astronomy tells us about the heavens, not about how to get to heaven. The discovery of life elsewhere would have an effect, though, on the views of the fundamentalists. For the same reasons that they do not accept that evolution occurs on Earth, the existence of life on other planets and the possibility that that life could evolve would threaten their views.

And, what about the implications of the discovery of extraterrestrial intelligent life? Although very unlikely, such a discovery would have truly profound results. Most interestingly, human civilization has just begun within the last half century to have the ability to detect extraterrestrial intelligence, through radio communications, for example. The odds that we would find another civilization that also had just developed the ability to communicate are negligibly small. This means that, if we do discover intelligent life, it almost certainly would be more advanced technologically than we are. Such a discovery would be humbling and inspiring to terrestrial civilization, but at the same time it might be demoralizing and threatening to discover life that is so much more advanced than us.

Although it is difficult to predict what the effects of such a discovery would be on human civilization, most people suspect that it would depend on the distance to the other civilization. If they were on the other side of the galaxy, the extreme distance would preclude any two-way communications or physical contact, and their existence might not affect us very much. On the other hand, if they were on a planet around a nearby star, two-way communications would be relatively straightforward (even if a response to a message might take a couple of decades) and interstellar travel could be possible. Humans might have a very different response to such a nearby and imminent interaction.

Finally, what would be the implications of searching for life and not finding any? We have the ability today to travel to Mars and to determine definitively if there is life there, even though it might take decades or longer to carry out such a search. We also could send a spacecraft to Europa and to Titan to search for life or for information of exobiological relevance. At the rate that technology is advancing, it might not be very long before we have the ability to detect Earth-like planets in orbit around other stars and possibly even to image them or to measure their properties and look for evidence of life. Suppose we carry out these detailed investigations and find no evidence of any life.

Such a result would have a dramatic effect on us. It would challenge our view of the origin and evolution of planets and of life. Clearly, it would suggest either that the chemical view of the origin of life was incorrect or that the likelihood of such an event actually occurring was much lower than we imagine. Of course, one can never prove conclusively that no other life in the galaxy or in the universe exists, especially without the ability to travel to other solar systems and do direct searches.

At the current rate of exploration of the solar system and of the universe, though, it could be decades before we know the answer even for our own solar system. In science today, a period of decades is so long as to be functionally equivalent to an infinite length of time; our views of the evolution of the universe no doubt will evolve substantially in that time. On the other hand, it is possible that the fossil evidence found within the martian meteorites could be demonstrated, at any time and to everyone's satisfaction, as being of non-terrestrial biological origin.

The ongoing search for life on other planets

What are we likely to find out in the near future as a result of the ongoing exploration of our own solar system and the search for other planets? The focus for searching for life will be on the exploration of our own solar system. This is a question mainly of logistics rather than interest. The search for extrasolar planets is ongoing, of course, but unlikely to result in the discovery of Earth-like planets within the next few years. Similarly, imaging of planets around other stars, or spectroscopic observations that will tell us about their atmospheric composition, is a long time off. Our own solar system, however, is accessible today. The fact that the planned exploration program might take a decade or more to implement is not so much a matter of technology or capability as one of expense and desire.

Mars occupies a special place NASA's planetary exploration program and, in fact, has for several decades. This is driven in part by the proximity of the red planet, the long history of telescopic observations, the evidence for seasonal atmospheric and surface changes and for long-term climate change, the diverse and varied geology of the surface, and the possibility that life could exist on or near the surface. At present, NASA plans on sending spacecraft to Mars at every opportunity (about every two years, when the Earth and Mars line up) for the next decade or more. The goal of the series of missions is to obtain information that might answer the question of whether Mars has ever had life.

Two spacecraft were launched successfully toward Mars in 1996 by the United States, an orbiter and a lander. The orbiter, the *Mars Global Surveyor* spacecraft, is designed to replace much of the science from the *Mars Observer* spacecraft that was lost just as it arrived at Mars in 1993. Although it will not focus on possible martian biology, it will be making a variety of measurements that will provide a global context for the geological history of the planet and may be relevant to possible biology. *Mars Global Surveyor* went into orbit around Mars in late 1997, and will begin its global mapping after its orbit is circularized in early 1999.

Measurements of the global topography and gravity, and a determination of the existence of an intrinsic magnetic field and a possible remnant field, will be relevant to the geophysics of the planet and to the history of the planet's interior. High-resolution images of the planet (with a resolution as small as about 1.5 m) will be combined with this information to provide insight into the geological history of Mars.

Measurements of the mid-infrared thermal emission will be used to determine the mineralogy of the surface, and to map out the occurrence of different minerals over the surface. These measurements will not observe possible life directly, but can be used to search for minerals that are of biological interest. In particular, minerals associated with hot springs or hydrothermal vents can point to locations where liquid water, and therefore life, might exist today or have existed in the recent past. These locations will be sites of obvious interest for future lander spacecraft that will search for life.

[Figure 18.1]
View of the surface of Mars obtained from the *Mars Pathfinder* spacecraft shortly after its landing on July 4, 1997. The alignment of some of the larger rocks suggests that they were deposited by the floods that washed over the surface 1 or 2 billion years ago. (NASA photo.)

The *Mars Pathfinder* spacecraft is a lander, designed to demonstrate the ability to do a low-cost science mission to the martian surface. It landed successfully on July 4, 1997, and carried out a tremendously successful mission before losing contact with Earth in the late summer. It did not have any experiments aboard that are directly related to the existence of life, but it did make some relevant observations. It had a multispectral imager that could determine the mineralogical composition of the soil and of the rocks present on the surface, and a meteorology experiment designed to tell us about the local environment. In addition, it contained the first martian rover, a vehicle about the size of a microwave oven that can move around the surface under control from the Earth. The rover included an instrument that can determine the elemental abundances of surface materials, so that the composition of the soil and of the various rocks on the surface can be determined.

[Figure 18.2]
Mars Pathfinder image of the rover vehicle *Sojourner*. The microwave-oven-sized rover was able to move around on the surface, image features close up, and determine the elemental composition of soil and rock samples. (NASA photo.)

The *Pathfinder* spacecraft landed in a region of Mars that had been covered by one of the catastrophic floods 1 or 2 billion years ago (see Chapter 8). The surface as seen from the lander contained rocks and debris that had been deposited by the floods, and then not substantially modified for more than a billion years (Figure 18.1). In addition, stereo images showed the presence of large, 20-m-wavelength undulations in the surface, presumably left behind by the devastating floods. The rover was successfully deployed (Figure 18.2), and managed to traverse a total distance of more than 150 m before contact was lost; included in the traverses were analyses of a dozen soil and rock samples.

Two U.S. spacecraft will be launched to Mars in 1998 as well, also an orbiter and a lander. These are the *Mars Surveyor '98* spacecraft. The orbiter will include an imager and an instrument that can measure the properties of the atmosphere for a Mars year in order to understand the seasonal behavior of the volatiles. The lander will land at high southern latitudes, on the deposits that surround the residual south polar ice cap. Its purpose is to examine this essentially unexplored territory on Mars, because the polar ice deposits are thought to play such an important role in the history of the atmosphere. Measurements will include the composition of the ice and of any non-icy material (such as dirt or rock) at the landing site, the physical nature of the surface ice itself, and perhaps the relative abundances of several of the stable isotopes that can tell us about the history of the atmosphere.

Japan also will launch a spacecraft to Mars in 1998, the *Planet-B* mission. It will investigate the nature of the upper atmosphere and the way in which the solar wind interacts with it. Because of the importance of escape of the upper atmosphere to space in the history of the martian climate, these measurements may be very important in piecing the history together.

Two spacecraft will be launched in 2001, also an orbiter and a lander. The orbiter will contain an instrument that can map out the elemental composition of the entire surface at a resolution of a few hundred kilometers. Included in these measurements will be a determination of the abundance of water near the martian surface (the top half-meter or so); this will allow global maps to be made of the distribution of ground ice, water of hydration of minerals, and possibly groundwater. Of particular relevance to exobiology, the orbiter will also contain an infrared spectrometer with ~100-m spatial resolution. This will be able to map the minerals of the surface, in order to look for small-scale deposits of minerals that would be indicative of the presence of hydrothermal systems or aqueous environments. Such regions will be of the most interest for detailed investigation by landers and rovers.

The lander will contain a rover vehicle that is supposed to be able to traverse a total distance of several tens of kilometers. This would allow it to get to some of these small-scale sites of interest and to collect samples from them. The rover will be about 1 m in length, and will contain various remote-sensing and *in situ* analysis instruments to examine rock and soil samples prior to collection.

Although not yet well defined, missions to be launched in 2003 may be very similar to those launched in 2001.

Finally, in 2005, a spacecraft will be launched to Mars that will land, pick up samples of rock and soil that had been collected by either the 2001 or 2003 rovers, and bring them back to the Earth. If the sample site is chosen carefully to be a site of exobiological relevance, such as a hydrothermal vent or a hot spring, we will be able to make a direct examination of materials of biological interest, including possible fossils of extinct martian life and even possible present-day life. Even if the best site for biology is not chosen, the samples will provide a tremendous amount of information about Mars. In addition to confirming the martian origin for the 'martian' meteorites, they would allow more detailed investigation of the history of the surface and atmosphere, and would provide substantial new constraints on the history of a possible biosphere.

Bringing martian samples that have the potential to contain active biological organisms back to the Earth raises some interesting issues about the possible contamination of the Earth's biosphere. Few people believe that there is any real risk of interaction between possible martian organisms and terrestrial organisms. However, prudence dictates that the samples be treated as if they were hazardous, just in case there might be an interaction. Samples would need to be brought back to the Earth in sealed containers and analyzed, at least initially, in an isolated environment. The necessary precautions are relatively straightforward, however, and we already have the technological capability to implement them properly.

There are plans and opportunities for missions that are pertinent to exobiology and that will go to places other than Mars, as well.

The *Cassini* mission was launched in 1997 to Saturn and to its satellite Titan. It includes an orbiter that will orbit Saturn and a descent vehicle that will probe Titan's atmosphere directly. Of special interest are the radar measurements that will be made of the Titan surface and the compositional information that will be obtained for the Titan atmosphere. The radar measurements will allow a global map of the surface to be made despite the clouds that otherwise obscure it. This will give us definitive information on whether a global ocean of methane and ethane exists on the surface, or on whether isolated lakes or even ponds of liquid exist. This information will have implications for understanding the long-term volatile history of the planet and the history of the atmospheric organic chemical processes. The measurements of the composition of the atmosphere will provide constraints on the formation, abundance, and evolution of organic compounds, and also on the history of the atmosphere and volatiles. Together, these new data will tell us much about Titan's environment and, by analogy, possibly about the evolution of the early Earth.

Although there are no definite plans to explore Europa, detailed investigation obviously would be relevant to possible Europan biology. Although an ocean on Europa would be buried far enough below the surface that it would not be easily accessed, there still are possibilities for exploration. There might be recently deposited water ice or active spring vents at the surface that could be sampled, and any recent activity might have brought any organic molecules or living organisms to the surface where they also could be sampled. One mission scenario includes landing a spacecraft onto the Europa surface to look for organic molecules; another would involve sending an impacting device that could hit the surface and throw organic debris up into space where it could be sampled directly by a flyby spacecraft; a third would utilize a melting device to allow a probe to descend through the ice cover to a buried ocean and to explore that ocean. Any of these approaches would begin to explore the possibilities for life on Europa.

Another key part of the search for life is the ongoing search for planets around other stars. Planets and brown dwarfs have been known only since about 1990, and the number of known planets is quite small – perhaps a dozen confirmed planets and a couple dozen that have not yet been confirmed. During the next decade, we can expect that the technical capability of the several approaches to looking for planets will improve dramatically. Finding Jupiter-like planets if they exist around the several hundred closest stars should be possible. Much better results may be achieved using Earth-orbiting spacecraft. Certainly, we can expect to have data on enough star systems (and any planets they might have) to have a better idea as to what is typical of planetary systems: Are Jupiter-like planets common? Does the distribution of planetary sizes, and their locations within their solar systems, match the expectations based on current theories? Finding Earth-like planets, though, is so much more difficult than finding gas giants that we cannot expect searches to be successful in the near future.

Finally, we should not overlook the significance of the ongoing search for clues to the origin and early evolution of life on Earth. Clearly, new results in this area would be of fundamental importance to understanding the possible occurrence of extraterrestrial life. What was the specific mechanism that allowed a transition to occur between non-living organic molecules and self-replicating and metabolizing entities on the early Earth? How simple (or how complex) were the first organisms that we would consider to be life? When and how did RNA become important for life? Was there an early RNA world, and how did it undergo a transition to the modern DNA world? These are questions that we may get answers to during the coming decade. Certainly, we will make substantial progress in understanding these issues, and any progress will help us in our search for life in the universe.

Concluding comments

It is tempting to try to say something grand and eloquent to close off the discussion regarding life in the universe. The subject is so significant and fundamental in its own right, however, that it speaks for itself. The questions surrounding whether we on Earth are the only forms of life in the universe or whether life is widespread throughout the galaxy are deeply profound. They involve nothing less than telling us how we fit into the universe around us. Within our lifetimes, we may have answers to some or all of these questions. Certainly, we will know much more about our origins than we know today.

We may have already turned the corner in our understanding of life in the universe, despite the fact that we have not found any direct evidence for the existence of life elsewhere. Many, if not most, scientists today believe that it is very likely that life exists on planets other than the Earth. Certainly, as we begin to explore the universe around us, we will find out more about the planets, their environments, and their occurrence around other stars. This information will provide us with the information that will help us to determine whether life does exist elsewhere or whether it does not. Despite all of our theories, however, the universe no doubt holds many surprises for us. We have to expect that many of the details discussed in this book will change in a steady stream of new discoveries; however, despite the changing specific details, we expect that the basics in our overall understanding of the universe, the planets, our Earth, and life are correct. As we explore the world around us, the search for life on other planets needs to be an experimental program, rather than a theoretical one. We need to continue our search for the evidence.

310

Chapter 1

Chang, S., Planetary environments and the conditions of life, *Phil. Trans. R. Soc. Lond. A*, **325**, 601–610, 1988.

Davies, P., *Are we alone? Philosophical implications of the discovery of extraterrestrial life*, Basic Books, N.Y., 1995.

Dick, S.J., *The biological universe – the twentieth-century extraterrestrial life debate and the limits of science*, Cambridge University Press, Cambridge, 1996.

Hartmann, W.K., *Moons and planets*, 3rd edn., Wadsworth Publ. Co., Belmont, CA, 1993.

Kirshner, R.P., The Earth's elements, *Sci. Amer.*, **271**, 58–65, 1994.

Lederberg, J., Exobiology: approaches to life beyond the Earth, *Science*, **132**, 393–400, 1960.

Peebles, P.J.E., D.N. Schramm, E.L. Turner, and R.G. Kron, The evolution of the Universe, *Sci. Amer.*, **271**, 52–57, 1994.

Sagan, C., The origin of life in a cosmic context, *Origins Life*, **5**, 497–505, 1974.

Sagan, C., The search for extraterrestrial life, *Sci. Amer.*, **271**, 92–99, 1994.

Sullivan, W., *We are not alone, The continuing search for extraterrestrial intelligence*, Revised edn., Penguin Books, 1994.

Weinberg, S., Life in the Universe, *Sci. Amer.*, **271**, 44–49, 1994.

Chapter 2

Alvarez, W. and F. Asaro, What caused the mass extinction? – An extraterrestrial impact, *Sci. Amer.*, **263**, 78–84, 1990.

Browne, M.W., New clues to agent of life's worst extinction, *New York Times*, December 15, 1992.

Chyba, C.F., T.C. Owen, and W.-H. Ip, Impact delivery of volatiles and organic molecules to Earth, in *Hazards due to comets and asteroids*, T. Gehrels, ed., University Arizona Press, Tucson, 9–58, 1994.

Courtillot, V.E., What caused the mass extinction? – A volcanic eruption, *Sci. Amer.*, **263**, 85–92, 1990.

Erwin, D.H., The mother of mass extinctions, *Sci. Amer.*, **275**, 72–78, 1996.

Maher, K.A. and D.J. Stevenson, Impact frustration of the origin of life, *Nature*, **331**, 612–614, 1988.

McLaren, D.J. and W.D. Goodfellow, Geological and biological consequences of giant impacts, *Annu. Rev. Earth Planet. Sci.*, **18**, 123–171, 1990.

Oberbeck, V.R. and G. Fogleman, Impacts and the origin of life, *Nature*, **339**, 434, 1989.

Oberbeck, V.R. and G. Fogleman, Impact constraints on the environment for chemical evolution and the continuity of life, *Origins Life Evol. Biosph.*, **20**, 181–195, 1990.

Renne, P.R., Z. Zichao, M.A. Richards, M.T. Black, and A.R. Basu, Synchrony and causal relations between Permian-Triassic boundary crises and Siberian flood volcanism, *Science*, **269**, 1413–1416, 1995.

Sleep, N.H., K.J. Zahnle, J.F. Kasting, and H.J. Morowitz, Annihilation of ecosystems by large asteroid impacts on the early Earth, *Nature*, **342**, 139–142, 1989.

Wetherill, G.W., Possible consequences of absence of 'Jupiters' in planetary systems, *Astrophys. Space Sci.*, **212**, 23–32, 1994.

Chapter 3

Allegre, C.J. and S.H. Schneider, The evolution of the Earth, *Sci. Amer.*, **271**, 66–75, 1994.

Calvert, A.J., E.W. Sawyer, W.J. Davis, and J.N. Ludden, Archaean subduction inferred from seismic images of a mantle suture in the Superior province, *Nature*, **375**, 670–674, 1995.

Kasting, J.F., Earth's early atmosphere, *Science*, **259**, 920–926, 1993.

Larson, R.L., The mid-Cretaceous superplume episode, *Sci. Amer.*, **272**, 82–86, 1995.

Monroe, J.S. and R. Wicander, *Physical geology – exploring the Earth*, 2nd edn., West Publ. Co., San Francisco, 1995.

Stacey, F.D., *Physics of the Earth*, 3rd edn., Brookfield Press, Brisbane, 1992.

Taylor, S.R. and S.M. McLennan, The evolution of continental crust, *Sci. Amer.*, **274**, 76–81, 1996.

Walker, J.C.G., *Evolution of the atmosphere*, Macmillian Publ. Co., New York, 1977.

Wicander, R. and J.S. Monroe, *Historical geology – evolution of the Earth and life through time*, 2nd edn., West Publ. Co., San Francisco, 1993.

Zindler, A. and S. Hart, Chemical geodynamics, *Annu. Rev. Earth Planet. Sci.*, **14**, 493–571, 1986.

Chapter 4

de Duve, C., The birth of complex cells, *Sci. Amer.*, **274**, 50–57, 1996.

Gould, S.J., *Wonderful life – The Burgess shale and the nature of history*, W.W. Norton and Co., New York, 1989.

Gould, S.J., The evolution of life on the Earth, *Sci. Amer.*, **271**, 84–91, 1994.

Holland, H.D., *The chemical evolution of the atmosphere and oceans*, Princeton University Press, New Jersey, 1984.

Kasting, J.F., Earth's early atmosphere, *Science*, **259**, 920–926, 1993.

Kasting, J.F. and T.P. Ackerman, Climatic consequences of very high carbon dioxide levels in the Earth's early atmosphere, *Science*, **234**, 1383–1385, 1986.

Mojzsis, S.J., G. Arrhenius, K.D. McKeegan, T.M. Harrison, A.P. Nutman, and C.R.L. Friend, Evidence for life on Earth before 3,800 million years ago, *Nature*, **384**, 55–59, 1996.

Runnegar, B., Evolution of the earliest animals, in *Major events in the history of life*, J.W. Schopf, ed., Jones and Bartlett Publishers, Boston, 65–93, 1992.

Schopf, J.W., ed., *Earth's earliest biosphere – its origin and evolution*, Princeton University Press, New Jersey, 1983.

Schopf, J.W., The oldest fossils and what they mean, *Major events in the history of life*, J.W. Schopf, ed., Jones and Bartlett Publishers, Boston, 29–63, 1992.

Schopf, J.W., ed., *Major events in the history of life*, Jones and Bartlett Publishers, Boston, 1992.

Shixing, Z. and C. Huineng, Megascopic multicellular organisms from the 1700-million-year-old Tuanshanzi formation in the Jixian area, North China, *Science*, **270**, 620–622, 1995.

Walker, J.C.G., *Evolution of the atmosphere*, Macmillian Publ. Co., New York, 1977.

Walker, J.C.G., Carbon dioxide on the early Earth, *Origins Life*, **16**, 117–127, 1985.

Walker, J.C.G., C. Klein, M. Schidlowski, J.W. Schopf, D.J. Stevenson, and M.R.

312   Walter, Environmental evolution of the Archaean-early Proterozoic Earth, in *Earth's earliest biosphere – its origin and evolution*, J.W. Schopf, ed., Princeton University Press, New Jersey, 260–290, 1983.

Wicander, R. and J.S. Monroe, *Historical geology – evolution of the earth and life through time*, 2nd edn., West Publ. Co., San Francisco, 1993.

Chapter 5

Baross, J.A. and S.E. Hoffman, Submarine hydrothermal vents and associated gradient environments as sites for the origin and evolution of life, *Origins Life*, **15**, 327–345, 1985.

Broad, W.J., Drillers find lost world of ancient microbes, *New York Times*, 4 October, B5–B9, 1994.

Brock, T.D., M.T. Madigan, J.M. Martinko, and J. Parker, *Biology of microorganisms*, 7th edn., Prentice Hall, Englewood, NJ, 1994.

Bult, C.J., and 39 others, Complete genome sequence of the methanogenic Archaeon, *Methanococcus jannaschii*, *Science*, **273**, 1058–1072, 1996.

Campbell, N.A., *Biology*, 3rd edn., Benjamin/Cummings Publ. Co., Redwood City, California, 1993.

Doolittle, R.F., D. Feng, S. Tsang, G. Cho, and E. Little, Determining divergence times of the major kingdoms of living organisms with a protein clock, *Science*, **271**, 470–477, 1996.

Fredrickson, J.K. and T.C. Onstott, Microbes deep inside the Earth, *Sci. Amer.*, **275**, 68–73, 1996.

Friedmann, E.I., Endolithic microorganisms in the Antarctic cold desert, *Science*, **215**, 1045–1052, 1982.

Gold, T., The deep, hot biosphere, *Proc. Natl. Acad. Sci.*, **89**, 6045–6049, 1992.

Iwabe, N., K. Kuma, M. Hasegawa, S. Osawa, and T. Miyata, Evolutionary relationship of archaebacteria, eubacteria, and eukaryotes inferred from phylogenetic trees of duplicated genes, *Proc. Natl. Acad. Sci.*, **86**, 9355–9359, 1989.

Jannasch, H.W., Microbial processes at deep sea hydrothermal vents, in *Hydrothermal processes at seafloor spreading centers*, P.A. Rona *et al.*, eds., Plenum Press, 677–709, 1983.

Madigan, M.T. and B.L. Marrs, Extremophiles, *Sci. Amer.*, **276**, 82–87, 1997.

Nealson, K.H. and D. Saffarini, Iron and manganese in anaerobic respiration: environmental significance, physiology, and regulation, *Annu. Rev. Microbiol.*, **48**, 311–343, 1994.

Parkes, R.J., B.A. Cragg, S.J. Bale, J.M. Getliff, K. Goodman, P.A. Rochelle, J.C. Fry, A.J. Weightman, and S.M. Harvey, Deep bacterial biosphere in Pacific Ocean sediments, *Nature*, **371**, 410–413, 1994.

Segerer, A.H., S. Burggraf, G. Fiala, G. Huber, R. Huber, U. Pley, and K.O. Stetter, Life in hot springs and hydrothermal vents, *Origins Life Evol. Biosph.*, **23**, 77–90, 1993.

Shock, E.L., T. McCollom, and M.D. Schulte, Geochemical constraints on chemolithoautotrophic reactions in hydrothermal systems, *Origins Life Evol. Biosph.*, **25**, 141–159, 1995.

Stevens, T.O., and J.P. McKinley, Lithoautotrophic microbial ecosystems in deep
basalt aquifers, *Science*, **270**, 450–454, 1995.

Von Damm, K.L., Seafloor hydrothermal activity: black smoker chemistry and
chimneys, *Annu. Rev. Earth Planet. Sci.*, **18**, 173–204, 1990.

Woese, C.R. Bacterial evolution, *Microbiol. Rev.*, **51**, 221–271, 1987.

Chapter 6

Baross, J.A. and S.E. Hoffman, Submarine hydrothermal vents and associated
gradient environments as sites for the origin and evolution of life, *Origins Life*,
15, 327–345, 1985.

Cairns-Smith, A.G., The first organisms, *Sci. Amer.*, **252**, 90–100, 1985.

Cech, T.R., A model for the RNA-catalyzed replication of RNA, *Proc. Natl. Acad. Sci.*,
83, 4360–4363, 1986.

Chang, S., D. DesMarais, R. Mack, S.L. Miller, and G.E. Strathearn, Prebiotic
organic syntheses and the origin of life, in *Earth's earliest biosphere – its origin
and evolution*, J.W. Schopf, ed., Princeton University Press, New Jersey, 53–92
1983.

Chyba, C. and G.D. McDonald, The origin of life in the solar system: current issues,
Annu. Rev. Earth Planet. Sci., **23**, 215–249, 1995.

Chyba, C. and C. Sagan, Endogenous production, exogenous delivery, and impact-
shock synthesis of organic molecules: an inventory for the origins of life,
Nature, **355**, 125–131, 1992.

Corliss, J.B., J.A. Baross, and S.E. Hoffman, An hypothesis concerning the
relationship between submarine hot springs and the origin of life on Earth,
Oceanol. Acta, **SP-8**, 59–69, 1981.

Ferris, J.P., Catalysis and prebiotic RNA synthesis, *Origins Life Evol. Biosph.*, **23**,
307–315, 1993.

Ferris, J.P., A.R. Hill, Jr., R. Liu, and L.E. Orgel, Synthesis of long prebiotic oligomers
on mineral surfaces, *Nature*, **381**, 59–61, 1996.

Hegstrom, R.A. and D.K. Kondepudi, The handedness of the universe, *Sci. Amer.*,
262, 108–115, 1990.

Holm, N.G., Why are hydrothermal systems proposed as plausible environments
for the origin of life? *Origins Life Evol. Biosph.*, **22**, 5–14, 1992.

Horgan, J., In the beginning ..., *Sci. Amer.*, **264**, 116–125, 1991.

Joyce, G.F., The rise and fall of the RNA world, *New Biologist*, **3**, 399–407, 1991.

Miller, S.L., A production of amino acids under possible primitive Earth
conditions, *Science*, **117**, 528–529, 1953.

Miller, S.L., The prebiotic synthesis of organic compounds as a step toward the
origin of life, in *Major events in the history of life*, J.W. Schopf, ed., Jones and
Bartlett Publishers, Boston, 1–28, 1992.

Orgel, L.E., The origin of life on the Earth, *Sci. Amer.*, **271**, 76–83, 1994.

Oro, J., S.L. Miller, and A. Lazcano, The origin and early evolution of life on Earth,
Annu. Rev. Earth Planet. Sci., **18**, 317–356, 1990.

Shock, E.L., Geochemical constraints on the origin of organic compounds in
hydrothermal systems, *Origins Life Evol. Biosph.*, **20**, 331–367, 1990.

Shock, E.L., T.McCollom, and M.D. Schulte, Geochemical constraints on

314   chemolithoautotrophic reactions in hydrothermal systems, *Origins Life Evol. Biosph.*, **25**, 141–159, 1995.

Chapter 7

Chang, S., Planetary environments and the conditions of life, *Phil. Trans. R. Soc. Lond. A*, **325**, 601–610, 1988.

Chyba, C. and G.D. McDonald, The origin of life in the solar system: current issues, *Annu. Rev. Earth Planet. Sci.*, **23**, 215–249, 1995.

Firsoff, V.A., Possible alternative chemistries to life, *Spaceflight*, **7**, 132–136, 1965.

Lederberg, J., Exobiology: approaches to life beyond the Earth, *Science*, **132**, 393–400, 1960.

Pimentel, G.C., K.C. Atwood, H. Gaffron, H.K. Hartline, T.H. Jukes, E.C. Pollard, and C. Sagan, Exotic biochemistries in exobiology, in *Biology and the exploration of Mars*, C.S. Pittendrigh, W. Vishniac, and J.P.T. Pearman, eds., Publ. 1296, National Research Council, Washington, DC, 243–251, 1966.

Sagan, C., The search for extraterrestrial life, *Sci. Amer.*, **271**, 92–99, 1994.

Wald, G., The origins of life, *Proc. Natl. Acad. Sci.*, **52**, 595–611, 1964.

Chapter 8

Boston, P.J., M.V. Ivanov, and C.P. McKay, On the possibility of chemosynthetic ecosystems in subsurface habitats on Mars, *Icarus*, **95**, 300–308, 1992.

Brakenridge, G.R., H.E. Newsom, and V.R. Baker, Ancient hot springs on Mars: origin and paleoenvironmental sigificance of small martian valleys, *Geology*, **13**, 859–862, 1985.

Carr, M.H., *The surface of Mars*, Yale University Press, New Haven, 232 pp., 1981.

Carr, M.H., *Water on Mars*, Oxford University Press, N.Y., 229 pp., 1996.

Clark, B.C., Solar-driven chemical energy source for a martian biota, *Origins Life*, **9**, 241–249, 1979.

Farmer, J., D. DesMarais, R. Greeley, R. Landheim, and H. Klein, Site selection for Mars exobiology, *Adv. Space Res.*, **15**, 157–162, 1995.

Gladman, B.J., J.A. Burns, M. Duncan, P. Lee, and H.F. Levison, The exchange of impact ejecta between terrestrial planets, *Science*, **271**, 1387–1392, 1996.

Jakosky, B.M. and R.M. Haberle, The seasonal behavior of water on Mars, in *Mars*, H.H. Kieffer, B.M. Jakosky, C.W. Snyder, and M.S. Matthews, eds., University Arizona Press, 969–1016, 1992.

Jakosky, B.M. and J.H. Jones, The history of martian volatiles, *Rev. Geophys.*, **35**, 1–16, 1997.

Lederberg, J. and C. Sagan, Microenvironments for life on Mars, *Proc. Natl. Acad. Sci.*, **48**, 1743–1745, 1962.

Kieffer, H.H., B.M. Jakosky, C.W. Snyder, and M.S. Matthews, eds., *Mars*, University Arizona Press, 1498 pp., 1992.

Klein, H.P., The Viking mission and the search for life on Mars, *Rev. Geophys. Space Phys.*, **17**, 1655–1662, 1979.

Klein, H.P., On the search for extant life on Mars, *Icarus*, **120**, 431–436, 1996.

Klein, H.P., N.H. Horowitz, and K. Biemann, The search for extant life on Mars, in Kieffer, H.H., B.M. Jakosky, C.W. Snyder, and M.S. Matthews, eds., *Mars*, University Arizona Press, 1221–1233, 1992.

McKay, C.P., R.L. Mancinelli, C.R. Stoker, and R.A. Wharton, Jr., The possibility
of life on Mars during a water-rich past, in *Mars*, H.H. Kieffer, B.M. Jakosky,
C.W. Snyder, and M.S. Matthews, eds., University Arizona Press, 1234–1245,
1992.

NASA Exobiology Program Office, An exobiological strategy for Mars exploration,
NASA SP-530, 56 pp, 1995.

Squyres, S.W. and J.F. Kasting, Early Mars: how warm and how wet? *Science*, **265**,
774–749, 1994.

Chapter 9

Anders, E., Evaluating the evidence for past life on Mars, *Science*, **274**, 2119–2121,
1996.

Ash, R.D., S.F. Knott, and G. Turner, A 4-Gyr shock age for a martian meteorite and
implications for the cratering history of Mars, *Nature*, **380**, 57–59, 1996.

Brock, T.D., M.T. Madigan, J.M. Martinko, and J. Parker, *Biology of microorganisms*,
7th edn., Prentice Hall, Englewood, NJ, 1994.

Campbell, N.A., *Biology*, 3rd edn., Benjamin/Cummings Publ. Co., Redwood City,
California, 1993.

Folk, R.L., SEM imagine of bacteria and nannobacteria in carbonate sediments
and rocks, *J. Sed., Petrol.*, **63**, 990–999, 1993.

Harvey, R.P. and H.Y. McSween Jr., A possible high-temperature origin for the
carbonates in the martian meteorite ALH84001, *Nature*, **382**, 49–51, 1996.

Kerridge, J.F., Origin of amino acids in the early solar system, *Adv. Space Res.*, **15**,
107–111, 1995.

McKay, D.S., E.K. Gibson Jr., K.L. Thomas-Keprta, H. Vali, C.S. Romanek, S.J.
Clemett, X.D.F. Chillier, C.R. Maechling, and R.N. Zare, Search for past life on
Mars: possible relic biogenic activity in martian meteorite ALH84001, *Science*,
273, 924–930, 1996.

Mittlefehldt, D.W., ALH84001, a cumulate orthopyroxenite member of the martian
meteorite clan, *Meteoritics*, **29**, 214–221, 1994.

NASA Exobiology Program Office, An exobiological strategy for Mars exploration,
NASA SP-530, 56 pp, 1995.

Romanek, C.S., M.M. Grady, I.P. Wright, D.W. Mittlefehldt, R.A. Sockl, C.T. Pillinger,
and E.K. Gibson Jr., Record of fluid-rock interactions on Mars from the
meteorite ALH84001, *Nature*, **372**, 655–657, 1994.

Wing, M.R. and J.L. Bada, The origin of the polycyclic aromatic hydrocarbons in
meteorites, *Origins Life Evol. Biosph.*, **21**, 375–383, 1992.

Chapter 10

Carlotto, M.J., Digital imagery analysis of unusual martian surface features [Face
on Mars], *Appl. Opt.*, **27**, 1926–1933, 1988.

Carr, M.H., *The surface of Mars*, Yale University Press, New Haven, 232 pp., 1981.

Di Pietro, V. and G. Molenaar, *Unusual martian surface features* [Face on Mars],
Mars Research, Glenn Dale, Maryland, 77 pp., 1982.

Haynes, R.H. and C.P. McKay, The implantation of life on Mars: feasibility and
motivation, *Adv. Space Res.*, **12**, 133–140, 1992.

McKay, C.P., Does Mars have rights? An approach to the environmental ethics of planetary engineering, in *Moral Expertise*, D. MacNiven, ed., Routledge, New York, 184–197, 1990.

McKay, C.P. and R.H. Haynes, Should we implant life on Mars? *Sci. Amer.*, **263**, 144, 1990.

McKay, C.P., O.B. Toon, and J.F. Kasting, Making Mars habitable, *Nature*, **352**, 489–496, 1991.

Pollack, J.B. and C. Sagan, Planetary engineering, in *Resources of near-Earth space*, J. Lewis, M.S. Matthews, and M.L. Guerrieri, eds., University Arizona Press, Tucson, 921–950, 1993.

Space Studies Board, National Research Council, *Biological contamination of Mars – issues and recommendations*, National Academy Press, Washington, DC, 115 pp., 1992.

Chapter 11

Bougher, S.W., D.M. Hunten, and R.J. Phillips, eds., *Venus II : geology, geophysics, atmosphere, and solar wind environment*, University Arizona Press, Tucson, 1997.

Bullock, M.A. and D.H. Grinspoon, The stability of climate on Venus, *J. Geophys. Res.*, **101**, 7521–7529, 1996.

Fegley, B., Jr., and A.H. Treiman, Chemistry of atmosphere-surface interactions on Venus and Mars, in *Venus and Mars: atmospheres, ionospheres, and solar wind interactions*, J.G. Luhmann, M. Tatrallyay, and R.O. Pepin, eds., Geophysical Monograph 66, Amer. Geophys. Union, 7–71, 1992.

Grinspoon, D.H., Implications of the high D/H ratio for the sources of water in Venus' atmosphere, *Nature*, **363**, 428–431, 1993.

Grinspoon, D.H., *Venus revealed*, Addison Wesley, 1997.

Head, J.W., L.S. Crumpler, J.C. Aubele, J.E. Guest and R.S. Saunders, Venus volcanism: classification of volcanic features and structures, associations, and global distribution from Magellan data, *J. Geophys. Res.*, **97**, 13 153–13 197, 1992.

Hunten, D.M., L. Colin, T.M. Donahue, and V.I. Moroz, eds., *Venus*, University Ariz. Press, 1143 pp., 1983.

Ingersoll, A.P., The runaway greenhouse: a history of water on Venus, *J. Atmos. Sci.*, **26**, 1191–1198, 1969.

Pollack, J.B. and C. Sagan, Planetary engineering, in *Resources of near-Earth space*, J. Lewis, M.S. Matthews, and M.L. Guerrieri, eds., University Arizona Press, Tucson, 921–950, 1993.

Pollack, J.B., O.B. Toon, and R. Boese, Greenhouse models of Venus' high surface temperatures, as constrained by Pioneer Venus measurements, *J. Geophys. Res.*, **85**, 8223–8231, 1980.

Schaber, G.G., R.G. Strom, H.J. Moore, L.A. Soderblom, R.L. Kirck, D.J. Chadwick, D.D. Dawson, L.R. Gaddis, J.M. Boyce, and J. Russell, Geology and distribution of impact craters on Venus: what are they telling us?, *J. Geophys. Res.*, **97**, 13 257–13 301, 1992.

Schubert, G. and D.T. Sandwell, A global survey of possible subduction sites on Venus, *Icarus*, **117**, 173–196, 1995.

Solomon, S.C., S.E. Smrekar, D.L. Bindschadler, R.E. Grimm, W.M. Kaula, G.E. **317**
McGill, R.J. Phillips, R.S. Saunders, G. Schubert, S.W. Squyres, and E.R. Stofan,
Venus tectonics: an overview of Magellan observations, *J. Geophys. Res.*, **97**,
13199–13255, 1992.

Turcotte, D.L., Magellan and comparative planetology, *J. Geophys. Res.*, **101**,
4765–4773, 1996.

Chapter 12

Lara, L.M., R.D. Lorenz, and R. Rodrigo, Liquids and solids on the surface of
Titan: results of a new photochemical model, *Planet. Space Sci.*, **42**, 5–14,
1994.

Lunine, J.I., Does Titan have an ocean? A review of current understanding of Titan's
surface, *Rev. Geophys.*, **31**, 133–149, 1993.

Lunine, J.I. and C.P. McKay, Surface-atmosphere interactions on Titan compared
with those on the prebiotic Earth, *Adv. Space Res.*, **15**, 303–311, 1995.

Lunine, J.I., D.J. Stevenson, and Y.L. Yung, Ethane ocean on Titan, *Science*, **222**,
1229–1230, 1983.

Morrison, D., T. Owen, and L. Soderblom, The satellites of Saturn, in *Satellites*, J. A.
Burns and M.S. Matthews, eds., University Arizona Press, 765–801, 1986.

Muhleman, D.O., A.W. Grossman, B.J. Butler, and M.A. Slade, Radar reflectivity of
Titan, *Science*, **248**, 975–980, 1990.

Sagan, C., W.R. Thompson, and B.N. Khare, Titan: a laboratory for prebiological
organic chemistry, *Accounts Chem. Res.*, **25**, 286–292, 1992.

Sagan, C., W.R. Thompson, and B.N. Khare, Titan's organic chemistry: results of
simulation experiments, in *Proc. Symp. on Titan*, ESA SP-338, 161–165, 1992.

Smith, P.H., M.T. Lemmon, R.D. Lorenz, L.A. Sromovsky, J.J. Caldwell, and M.D.
Allison, Titan's surface, revealed by HST imaging, *Icarus*, **119**, 336–349, 1996.

Stevenson, D.J., The interior of Titan, in *Proc. Symp. on Titan*, ESA SP-338, 29–33,
1992.

Thompson, W.R. and C. Sagan, Organic chemistry on Titan – surface interactions,
in *Proc. Symp. on Titan*, ESA SP-338, 167–176, 1992.

Chapter 13

Ingersoll, A.P., Jupiter and Saturn, in *The solar system – observations and
interpretation*, M.G. Kivelson, ed., Prentice-Hall, Inc., New Jersey, 230–253,
1986.

Malin, M.C. and D.C. Pieri, Europa, in *Satellites*, J.A. Burns and M.S. Matthews,
eds., University Arizona Press, Tucson, 689–717, 1986

McDonald, G.D., B.N. Khare, and C. Sagan, Synthesis of organic macromolecules
in the Jovian atmosphere, *Abstracts for Division of Planetary Sciences Meeting*,
Hawaii, 87, 1995.

Nash, D.B., M.H. Carr, J. Gradie, D.M. Hunten, and C.F. Yoder, Io, in *Satellites*, J.A.
Burns and M.S. Matthews, eds., University Arizona Press, Tucson, 629–688,
1986.

Niemann, H.B., S.K. Atreya, G.R. Carignan, T.M. Donahue, J.A. Haberman, D.N.
Harpold, R.E. Hartle, D.M. Hunten, W.T. Kasprzak, P.R. Mahaffy, T.C. Owen,

N.W. Spencer, and S.H. Way, The Galileo Probe mass spectrometer: composition of Jupiter's atmosphere, *Science*, **272**, 846–849, 1996.

Ojakangas, G.W. and D.J. Stevenson, Thermal state of an ice shell on Europa, *Icarus*, **81**, 220–241, 1989.

Reynolds, R.T., S.W. Squyres, D.S. Colburn, and C.P. McKay, On the habitability of Europa, *Icarus*, **56**, 246–254, 1983.

Sagan, C. and E.E. Salpeter, Particles, environments, and possible ecologies in the Jovian atmosphere, *Astrophys. J. Suppl.*, **32**, 737–755, 1976.

Schubert, G., T. Spohn, and R.T. Reynolds, Thermal histories, compositions, and internal structures of the moons of the solar system, in *Satellites*, J.A. Burns and M.S. Matthews, eds., University Arizona Press, Tucson, 224–292, 1986.

Squyres, S.W., R.T. Reynolds, P.M. Cassen, and S.J. Peale, Liquid water and active resurfacing on Europa, *Nature*, **301**, 225–226, 1983.

Young, R.E., M.A. Smith, and C.K. Sobeck, Galileo Probe: in situ observations of Jupiter's atmosphere, *Science*, **272**, 837–838, 1996.

Chapter 14

Beckwith, S.V.W. and A.I. Sargent, Circumstellar disks and the search for neighbouring planetary systems, *Nature*, **383**, 139–144, 1996.

Boss, A.P., Extrasolar planets, *Phys. Today*, **49**, 32–38, 1996.

Chaisson, E. and S. McMillan, *Astronomy today*, Prentice Hall, New Jersey, 1996.

Kerridge, J.F., What can meteorites tell us about nebular conditions and processes during planetesimal accretion? *Icarus*, **106**, 135–150, 1993.

Lada, C.J. and F.H. Shu, The formation of sunlike stars, *Science*, **248**, 564–572, 1990.

Lissauer, J.J., Planet formation, *Annu. Rev. Astron. Astrophys.*, **31**, 129–174, 1993.

Lissauer, J.J., J.B. Pollack, G.W. Wetherill, and D.J. Stevenson, Formation of the Neptune system, in *Neptune and Triton*, D.P. Cruikshank, ed., University Arizona Press, Tucson, 37–108, 1995.

Mumma, M.J., P.R. Weissman, and S.A. Stern, Comets and the origin of the solar system: reading the Rosetta stone, in *Protostars and planets III*, ed. by E.H. Levy and J.I. Lunine, University Arizona Press, Tucson, 1177–1252,1993.

Wetherill, G.W., Formation of the Earth, *Annu. Rev. Earth Planet. Sci.*, **18**, 205–256, 1990.

Wetherill, G.W., Occurrence of Earth-like bodies in planetary systems, *Science*, **253**, 535–538, 1991.

Wetherill, G.W., Provenance of the terrestrial planets, *Geochim. Cosmochim. Acta*, **58**, 4513–4520, 1994.

Wetherill, G.W., How special is Jupiter? *Nature*, **373**, 470, 1995.

Wetherill, G.W., The formation and habitability of extra-solar planets, *Icarus*, **119**, 219–238, 1996.

Chapter 15

Beckwith, S.V.W. and A.I. Sargent, Circumstellar disks and the search for neighbouring planetary systems, *Nature*, **383**, 139–144, 1996.

Boss, A.P., Proximity of Jupiter-like planets to low-mass stars, *Science*, **267**, 360–362, 1995.

Boss, A.P., Extrasolar planets, *Phys. Today*, **49**, 32–38, 1996.

Butler, R.P. and G.W. Marcy, A planet orbiting 47 Ursae Majoris, *Astrophys. J.*, **464**, **319**
L153–L156, 1996.

Chaisson, E. and S. McMillan, *Astronomy today*, Prentice Hall, New Jersey, 1996.

Cochran, W.D., A.P. Hatzes, and T.J. Hancock, Constraints on the companion object to HD 114762, *Astrophys. J.*, **380**, L35–L38, 1991.

Lada, C.J. and F.H. Shu, The formation of sunlike stars, *Science*, **248**, 564–572, 1990.

Latham, D.W., T.Mazeh, R.P. Stefanik, M. Mayor, and G. Burki, The unseen companion of HD114762: a probable brown dwarf, *Nature*, **339**, 38–40, 1989.

Lin, D.N.C., P. Bodenheimer, and D.C. Richardson, Orbital migration of the planetary companion of 51 Pegasi to its present location, *Nature*, **380**, 606–607, 1996.

Lissauer, J.J., Planet formation, *Annu. Rev. Astron. Astrophys.*, **31**, 129–174, 1993.

Marcy, G.W. and R.P. Butler, A planetary companion to 70 Virginis, *Astrophy. J.*, **464**, L147–L151, 1996.

Mayor, M. and D. Queloz, A Jupiter-mass companion to a solar-type star, *Nature*, **378**, 355–359, 1995.

Nakajima, T., B.R. Oppenheimer, S.R. Kulkarni, D.A. Gollmowski, K. Matthews, and S.T. Durrance, Discovery of a cool brown dwarf, *Nature*, **378**, 463–465, 1995.

Oppenheimer, B.R., S.R. Kulkarni, K. Matthews, and T. Nakajima, Infrared spectrum of the cool brown dwarf Gl229B, *Science*, **270**, 1478–1479, 1995.

Rasio, F.A. and E.B. Ford, Dynamical instabilities and the formation of extrasolar planetary systems, *Science*, **274**, 954–956, 1996.

Wolszczan, A., Confirmation of Earth-mass planets orbiting the millisecond pulsar PSR B1257+12, *Science*, **264**, 538–542, 1994.

Chapter 16

Bullock, M.A. and D.H. Grinspoon, The stability of climate on Venus, *J. Geophys. Res.*, **101**, 7521–7529, 1996.

Hunten, D.M., Atmospheric evolution of the terrestrial planets, *Science*, **259**, 915–920, 1993.

Jakosky, B.M. and J.H. Jones, The history of martian volatiles, *Rev. Geophys.*, **35**, 1–16, 1997.

Kasting, J.F., D.P. Whitmire, and R.T. Reynolds, Habitable zones around main sequence stars, *Icarus*, **101**, 108–128, 1993.

Rampino, M.R. and K. Caldeira, The Goldilocks problem: climatic evolution and long-term habitability of terrestrial planets, *Annu. Rev. Astron. Astrophys.*, **32**, 83–114, 1994.

Reynolds, R.T., C.P. McKay, and J.F. Kasting, Europa, tidally heated oceans, and habitable zones around giant planets, *Adv. Space Res.*, **7**, 125–132, 1987.

Squyres, S.W. and J.F. Kasting, Early Mars: how warm and how wet? *Science*, **265**, 774–749, 1994.

Sundquist, E.T., The global carbon dioxide budget, *Science*, **259**, 934–941, 1993.

Wetherill, G.W., The formation and habitability of extra-solar planets, *Icarus*, **119**, 219–238, 1996.

Williams, D.M., J.F. Kasting, and R.A. Wade, Habitable moons around extrasolar giant planets, *Nature*, **385**, 234–236, 1997.

320 **Chapter 17**

Bracewell, R.N., *The galactic club: intelligent life in outer space*, W.H. Freeman and Co., San Francisco, 1974.

Calvin, W.H., The emergence of intelligence, *Sci. Amer.*, **271**, 100–107, 1994.

Cocconi, G. and P. Morrison, Searching for interstellar communications, *Nature*, **184**, 844–846, 1959.

Davies, P., *Are we alone? Philosophical implications of the discovery of extraterrestrial life*, Basic Books, N.Y., 1995.

Deardorff, J.W., Possible extraterrestrial strategy for Earth, *Q. J. R. Astron. Soc.*, **27**, 94–101, 1986.

Dick, S.J., Consequences of success in SETI: lessons from the history of science, in *Progress in the search for extraterrestrial life*, G.S. Shostak, ed., Astronomical Society of the Pacific, San Francisco, 521–532, 1995.

Dick, S.J., *The biological universe – the twentieth-century extraterrestrial life debate and the limits of science*, Cambridge University Press, Cambridge, 1996.

Gould, S.J., The evolution of life on the Earth, *Sci. Amer.*, **271**, 84–91, 1994.

Gould, S.J., *Full house: the spread of excellence from Plato to Darwin*, Crown Publishers, New York, 1996.

Hart, M.H., An explanation for the absence of extraterrestrials on Earth, in *Extraterrestrials: where are they?*, 2nd edn., B. Zuckerman and M.H. Hart, eds., Cambridge University Press, Cambridge, 1–8, 1995.

Mayr, E., The search for extraterrestrial intelligence, in *Extraterrestrials: where are they?*, 2nd edn., B. Zuckerman and M.H. Hart, eds., Cambridge University Press, Cambridge, 152–156, 1995.

Sagan, C., *The dragons of Eden – speculations on the evolution of human intelligence*, Random House, New York, 1977.

Sagan, C., The search for extraterrestrial life, *Sci. Amer.*, **271**, 92–99, 1994.

Sagan, C. and F. Drake, The search for extraterrestrial intelligence, *Sci. Amer.*, **232**, 80–89, 1975.

Shklovskii, I.S. and C. Sagan, *Intelligent life in the universe*, Holden-Day, San Francisco, 1966.

Sullivan, W., *We are not alone, the continuing search for extraterrestrial intelligence*, revised edn., Penguin Books, 1994.

Tarter, J., One attempt to find where they are: NASA's high resolution microwave survey, in *Extraterrestrials: where are they?*, 2nd edn., B. Zuckerman and M.H. Hart, eds., Cambridge University Press, Cambridge, 9–19, 1995.

Tattersall, I., Out of Africa again ... and again? *Sci. Amer.*, **276**, 60–67, 1997.

Wicander, R. and J.S. Monroe, *Historical geology – evolution of the Earth and life through time*, 2nd edn., West Publ. Co., San Francisco, 1993.

Chapter 18

Davies, P., *Are we alone? Philosophical implications of the discovery of extraterrestrial life*, Basic Books, N.Y., 1995.

Dick, S.J., *The biological universe – the twentieth-century extraterrestrial life debate and the limits of science*, Cambridge University Press, Cambridge, 1996.

National Academy of Sciences, Task Group on Issues in Sample Return, *Mars sample return – issues and recommendations*, National Research Council, Washington, DC, 1997.

National Academy of Sciences, *The search for life's origins – progress and future directions in planetary biology and chemical evolution*, National Research Council, Washington, DC, 1990.

Sullivan, W., *We are not alone, the continuing search for extraterrestrial intelligence*, revised edn., Penguin Books, 1994.

Van Doren, C., *A history of knowledge*, Ballantine Books, N.Y., 1991.

Figure references

322 Beckwith, S.V.W. and A.I. Sargent, Circumstellar disks and the search for
     neighbouring planetary systems, *Nature*, **383**, 139–144, 1996.

Boss, A.P., Proximity of Jupiter-like planets to low-mass stars, *Science*, **267**,
     360–362, 1995.

Brock, T.D., M.T. Madigan, J.M. Martinko, and J. Parker, *Biology of microorganisms*,
     7th edn., Prentice Hall, Englewood, NJ, 1994.

Cochran, W.D., A.P. Hatzes, and T.J. Hancock, Constraints on the companion object
     to HD 114762, *Astrophys. J.*, **380**, L35–L38, 1991.

Covey, C., S.L. Thompson, P.R. Weissman, and M.C. MacCracken, Global climatic
     effects of atmospheric dust from an asteroid or comet impact on Earth, *Global
     Planet. Change*, **9**, 263–273, 1994.

Fegley, B., Jr., and A.H. Treiman, Chemistry of atmosphere-surface interactions on
     Venus and Mars, in *Venus and Mars: atmospheres, ionospheres, and solar wind
     interactions*, J.G. Luhmann, M. Tatrallyay and R.O. Pepin, eds., Geophysical
     Monograph 66, Amer. Geophys. Union, 7–71, 1992.

Grieve, R.A. and E.M. Shoemaker, The record of past impacts on Earth, in *Hazards
     due to comets and asteroids*, T. Gehrels, ed., University Arizona Press, 417–462,
     1994.

Haberle, R.M., D. Tyler, C.P. McKay, and W.L. Davis, A model for the evolution of
     CO_2 on Mars, *Icarus*, **109**, 102–120, 1994.

Karlsson, H.R., R.N. Clayton, E.K. Gibson, Jr., and T.K. Mayeda, Water in SNC
     meteorites: evidence for a martian hydrosphere, *Science*, **255**, 1409–1411,
     1992.

Kasting, J.F., CO_2 condensation and the climate of early Mars, *Icarus*, **94**, 1–13,
     1991.

Kasting, J.F., D.P. Whitmire, and R.T. Reynolds, Habitable zones around main
     sequence stars, *Icarus*, **101**, 108–128, 1993.

Levin, G.V. and P.A. Straat, Recent results from the Viking Labelled Release
     experiment on Mars, *J. Geophys. Res.*, **82**, 4663–4667, 1977.

Lindal, G.F., G.E. Wood, H.B. Hotz, and D.N. Sweetnam, The atmosphere of Titan:
     an analysis of the Voyager 1 radio occultation measurements, *Icarus*, **53**,
     348–363, 1983.

Lunine, J.I., Titan's surface: nature and implications for Cassini, *Proceedings of the
     conference on the atmospheres of Saturn and Titan*, Alpbach, Austria, 1–6, 1985.

Lunine, J.I., Does Titan have an ocean? A review of current understanding of Titan's
     surface, *Rev. Geophys.*, **31**, 133–149, 1993.

Macdonald, G.A., A.A. Abbott, and F.L. Peterson, *Volcanoes in the sea – the geology
     of Hawaii*, 2nd edn., University of Hawaii Press, 1983.

Marcy, G.W. and R.P. Butler, A planetary companion to 70 Virginis, *Astrophy. J.*, **464**,
     L147–L151, 1996.

Marcy, G. and P. Butler, Searching for extrasolar planets, internet web site
     http://www.physics.sfsu.edu/~williams/planetsearch/planetsearch.html,
     1997.

McGill, G.E., J.L. Warner, M.C. Malin, R.E. Arvidson, E. Eliason, S. Nozette, and R.D.
     Reasenberg, Topography, surface properties, and tectonic evolution, in *Venus*,
     D.M. Hunten, L. Colin, T.M. Donahue, and V.I. Moroz, eds., University of
     Arizona Press, Tucson, 69–130, 1983.

McKay, D.S., E.K. Gibson Jr., K.L. Thomas-Keprta, H. Vali, C.S. Romanek, S.J. Clemett, X.D.F. Chillier, C.R. Maechling, and R.N. Zare, Search for past life on Mars: possible relic biogenic activity in martian meteorite ALH84001, *Science*, **273**, 924–930, 1996.

Melosh, H.J., and A.M. Vickery, Impact erosion of the primoridal atmosphere of Mars, *Nature*, **338**, 487–489, 1989.

Monroe, J.S. and R. Wicander, *Physical geology – exploring the Earth*, 2nd edn., West Publ. Co., San Francisco, 1995.

Morrison, D., C.R. Chapman, and P. Slovic, The impact hazard, in *Hazards due to comets and asteroids*, T. Gehrels, ed., University Arizona Press, 59–91, 1994.

Muhleman, D.O., A.W. Grossman, B.J. Butler, and M.A. Slade, Radar reflectivity of Titan, *Science*, **248**, 975–980, 1990.

Oppenheimer, B.R., S.R. Kulkarni, K. Matthews, and T. Nakajima, Infrared spectrum of the cool brown dwarf Gl229B, *Science*, **270**, 1478–1479, 1995.

Pepin, R.O., Meteorites – evidence of martian origin, *Nature*, **317**, 473–475, 1994.

Rampino, M.R. and B.M. Haggerty, Extraterrestrial impacts and mass extinctions of life, in *Hazards due to comets and asteroids*, T. Gehrels, ed., University Arizona Press, 827–857, 1994.

Sagan, C., *The dragons of Eden – speculations on the evolution of human intelligence*, Random House, New York, 1977.

Sagan, C. and F. Drake, The search for extraterrestrial intelligence, *Sci. Amer.*, **232**, 80–89, 1975.

Samuelson, R.E., W.C. Maguire, R.A. Hanel, V.G. Kunde, D.E. Jennings, Y.L. Yung, and A.C. Aikin, CO_2 on Titan, *J. Geophys. Res.*, **88**, 8709–8715, 1983.

Schaber, G.G., R.G. Strom, H.J. Moore, L.A. Soderblom, R.L. Kirck, D.J. Chadwick, D.D. Dawson, L.R. Gaddis, J.M. Boyce, and J. Russell, Geology and distribution of impact craters on Venus: what are they telling us?, *J. Geophys. Res.*, **97**, 13 257–13 301, 1992.

Schopf, J.W., The oldest fossils and what they mean, *Major events in the history of life*, J.W. Schopf, ed., Jones and Bartlett Publishers, Boston, 29–63, 1992.

Schubert, G. and D.T. Sandwell, A global survey of possible subduction sites on Venus, *Icarus*, **117**, 173–196, 1995.

Seiff, A., Thermal structure of the atmosphere of Venus, in *Venus*, D.M. Hunten, L. Colin, T.M. Donahue, and V.I. Moroz, eds., University of Arizona Press, Tucson, 215–279, 1983.

Shock, E.L., Chemical environments of submarine hydrothermal systems, *Origins Life Evol. Biosph.*, **22**, 67–107, 1992.

Tanaka, K.L., The stratigraphy of Mars, *J. Geophys. Res.*, **91**, E139–E158, 1986.

Wachtershauser, G., Before enzymes and templates: theory of surface metabolism, *Microbiol. Rev.*, **52**, 452–484, 1988.

Wetherill, G.W., Occurrence of Earth-like bodies in planetary systems, *Science*, **253**, 535–538, 1991.

Wetherill, G.W., The formation and habitability of extra-solar planets, *Icarus*, **119**, 219–238, 1996.

Wing, M.R. and J.L. Bada, The origin of the polycyclic aromatic hydrocarbons in meteorites, *Origins Life Evol. Biosph.*, **21**, 375–383, 1992.